SERIES INTRODUCTION

The Tutorial Texts series was begun in response to requests for copies of SPIE short course notes by those who were not able to attend a course. By policy the notes are the property of the instructors and are not available for sale. Since short course notes are intended only to guide the discussion, supplement the presentation, and relieve the lecturer of generating complicated graphics on the spot, they cannot substitute for a text. As one who has evaluated many sets of course notes for possible use in this series, I have found that material unsupported by the lecture is not very useful. The notes provide more frustration than illumination.

What the Tutorial Texts series does is to fill in the gaps, establish the continuity, and clarify the arguments that can only be glimpsed in the notes. When topics are evaluated for this series, the paramount concern in determining whether to proceed with the project is whether it effectively addresses the basic concepts of the topic. Each manuscript is reviewed at the initial state when the material is in the form of notes and then later at the final draft. Always, the text is evaluated to ensure that it presents sufficient theory to build a basic understanding and then uses this understanding to give the reader a practical working knowledge of the topic. References are included as an essential part of each text for the reader requiring more in-depth study.

One advantage of the Tutorial Texts series is our ability to cover new fields as they are developing. In fields such as sensor fusion, morphological image processing, and digital compression techniques, the textbooks on these topics were limited or unavailable. Since 1989 the Tutorial Texts have provided an introduction to those seeking to understand these and other equally exciting technologies. We have expanded the series beyond topics covered by the short course program to encompass contributions from experts in their field who can write with authority and clarity at an introductory level. The emphasis is always on the tutorial nature of the text. It is my hope that over the next few years there will be as many additional titles with the quality and breadth of the first seven years.

Donald C. O'Shea
Georgia Institute of Technology

CONTENTS

Preface

Mathematics is often called the queen of the sciences. Radiometry should then be called the waiting maid or servant. It is not especially elegant; it is not very popular, has not been trendy; but it is essential in almost every part of optical engineering. The infrared devices I described in *Introduction to Infrared System Design* (SPIE Press, 1996) cannot be designed without an understanding of the amount of power that impinges on the detector from the target, and the radiation of the target cannot be understood without a radiometric measurement. Similar statements can be made about photography, movies, TV, medical instrumentation, and industrial quality control.

Radiometry appears to be a simple subject, and the basics are indeed uncomplicated. The devil is in the details. These details include the language, which is special, but is essential for good understanding. Units and dimensions need to be followed assiduously, and everything must be considered for its pertinence and significance in measurements and calculations.

I have tried to address these issues in this text by discussing nomenclature in Chapter 2, by listing a taxonomy of measurements in Chapter 11, and by describing the most important techniques of measuring the major radiometric properties of flux and materials.

This text is the result of notes that I have used for an SPIE tutorial and from a three-hour course I have taught at the University of Arizona for a quarter century.

I am indebted to George Zissis, for whom I first worked in radiometry at The University of Michigan, and to Don O'Shea and Jim Palmer for their incisive criticisms of this text. I thank Eric Pepper, who thoroughly edited the text and caught a number of errors. The remaining misteaks are mein!

Of course I offer my gratitude to my wife of almost 45 years, who now knows what I do in front of the computer, and what I do with her telephone—and puts up with it.

I dedicate this book to the memory of my parents, who would have been pleased.

William L. Wolfe
March 1998

CHAPTER I

INTRODUCTION

Radiometry is an essential part of the optical design of almost every optical instrument. Such instruments are usually used to focus and detect radiation for some particular purpose, and for many applications it is absolutely essential to know how much radiation gets to the detector array or film in the image plane and the value of the resultant signal-to-noise ratio or exposure. Radiometry is almost essential in another sense, the measurement of the radiation of various objects. In fact, the word "radiometry" itself *means* the measurement of radiation. One cannot make the above-mentioned calculations without a knowledge of the flux from the source, whether it be a tungsten bulb or a sun-illuminated vista. Therefore, this text on radiometry involves both the techniques of calculating radiative transfer and the measurement of fluxes and radiometric properties of different sorts.

1.1 HISTORY

The most primitive beginnings of radiometry must have been the observation by early man of the different brightnesses of stars and the sensing of the warmth from the sun and the fire (after he invented or discovered it). These were radiometric measurements, but they surely were not quantitative. Greek astronomers, especially Ptolemy and Hipparchus, made good estimates of star magnitudes, and these were extended by Galileo[1].

The history of *quantitative* radiometry surely begins with the practice of photometry, the measurement of visible light. It was first put on an organized basis by Pierre Bouguer in 1729 when he described an instrument that could compare the brightnesses of two sources[2]. In 1760 Johann Lambert[3] enunciated

---

[1]Heath, T., *Greek Astronomy*, Library of Greek Thought, 147, 1932; T. S. Kuhn, *The Copernican Revolution,* Random House, 1959; G. Galileo, Dialogue Concerning the Two Chief World Systems—Ptolmaic and Copernican, University of California Press, 1967.

[2]Bouguer, P., Reprinted in Les Maítres de la Pensée Scientific, Paris, 1729; Histoire de l'Academe Royale, des Sciences, Paris 1726.

[3]Lambert, J. H., *Photometria sive de mensura et gradibus luminus colorum et umbrae*, 1760; a German translation by E. Anding in Ostwald, *Klassiker der exakten Wissenschaften,* Engelman, 1892.

the law of the addition of illumination, the inverse square law, cosine law of flux density distribution, and others. Many relatively small advances were made until Becquerel observed the photoelectric effect in 1839[4]. Then the photoconductive effect was discovered by Willoughby Smith in 1873 and the photoemissive effect by Hertz in 1887. These discoveries moved the different measurement instruments out of the realm of human observation and into that of quantitative analyses. Although the eye is a wonderful comparison device, it is notoriously poor at measuring radiation levels. Additional history is available in the very nice photometry text by Walsh[5] and the one on absolute radiometry by Hengstberger[6].

1.2 ORGANIZATION

The scheme of this text is to discuss first some of the basic concepts, the language of radiometry, and relatively simple radiative transfer. The methods used to describe the properties of radiators, reflectors, and transmitters are next described. Then ideal and practical sources are covered. Normalization, an interesting part of the field, involving making measurements and calculations of radiative quantities based on responses of certain sensors, is covered next. Radiometric standards are then described, just before measurement and calibration techniques. Radiometric, sometimes known as fake, temperatures are described along with their errors. Concepts of polarization are then discussed, mostly in terms of warnings.

[4]Hudson, R., *Infrared System Engineering*, Wiley, 1969.

[5]Walsh, J. W. T., *Photometry*, Third Edition, Dover, 1958.

[6]Hengstberger, E., ed., *Absolute Radiometry*, Academic Press, 1989.

CHAPTER 2

RADIOMETRIC QUANTITIES, THE LANGUAGE

Radiative transfer describes the ways that real bodies radiate power or energy to and from each other. This radiation is described in terms of energy, power, or certain geometric characteristics of power—its areal density (power per unit area), for example. An understanding of this transfer can only be obtained if the language is understood. So, at this point, several definitions are in order.

2.1 ANGLE AND SOLID ANGLE

The angle, more precisely the linear angle, is defined as the length of arc of a circle divided by the radius of the circle. An alternative and completely equivalent definition is the length of arc of a unit circle. The units of angle are degrees and radians and their subdivisions are arcminutes, arcseconds, milliradians, and microradians. It was the ancient Babylonians who first defined the degree as being $1/360^{th}$ of a full circle. The natural measure, however, is the radian. It is a length of arc divided by the radius, so that there are 2π radians in a full circle. But an angle is a subtense, a linear dimension divided by another linear dimension. As Figure 2-1 shows, a straight line and even a curved line can subtend the same angle as an arc on the circle. So, the definition should read that a linear angle is the projection of a line on a unit circle, and the line need not be straight. The angle is really determined by the endpoints of the line.

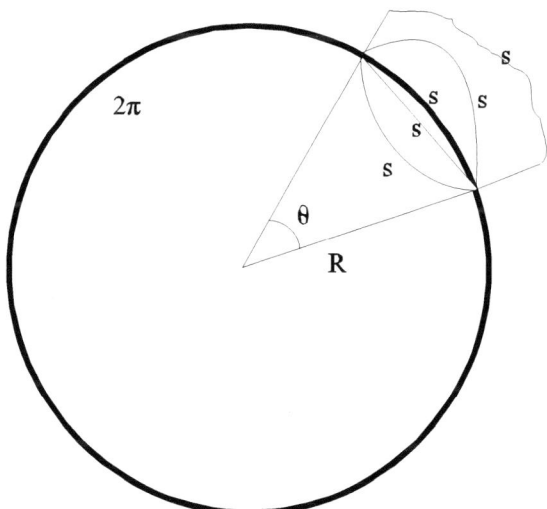

Figure 2-1. Illustration of linear angles, $\theta = s/R$, any s. Circumference $= 2\pi R$.

A solid angle is the two-dimensional equivalent of a linear angle. It is the projection of an area onto a unit sphere, as shown in Figure 2-2. The solid angle is determined by the perimeter of the area. An equivalent definition of a solid angle is the projection of an area onto a sphere divided by the square of the radius of the sphere. A sphere subtends 4π steradians; a hemisphere subtends half that. Square degrees or square arcseconds can be used, but steradians (from the Greek *stereos* for solid) are more appropriate and used by most people.

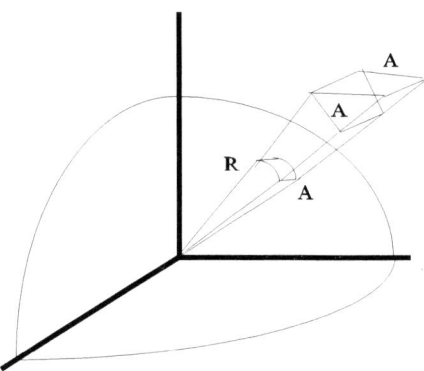

Figure 2-2. Illustration of a solid angle, $\Omega = A\cos\theta/R^2$, any A. Sphere area $= 4\pi R^2$.

2.2 PROJECTED AREA AND PROJECTED SOLID ANGLE

The idea of a projected area arises when an area in question is tilted with respect to the line of sight. As shown in Figure 2-3, an area that is tipped looks like a smaller area at the point of the observer. It will therefore either receive less flux or emit less flux to or from the observer. This is taken into account by using the projected area, which is the area times the cosine of the angle between the normal to the area and the line of sight, that is,

$$A_p = A\cos\theta, \tag{2-1}$$

where A_p is the projected area, A is the true, geometric area, and θ is the angle from the surface normal at which the area is viewed.

The projected solid angle is less straightforward. The solid angle is of itself calculated as the area projected onto the sphere, the center of which is the origin of flux, divided by the square of the distance from that center. This means that the solid angle is $A \cos\theta / \rho^2$, where A is the true area, θ is the angle between the line of sight and the normal to the surface, and ρ is the line-of-sight distance. The projected solid angle has one more cosine, that is,

$$\Omega_p = \Omega\cos\theta_1 = A\cos\theta_1\cos\theta_2, \qquad\qquad (2\text{-}2)$$

where Ω_p is the projected solid angle, Ω is the solid angle, and the two angles are with respect to the first area and the second area between which the radiation flows.

Normal Plate

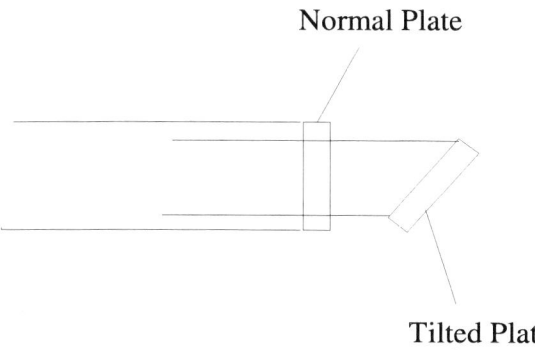

Tilted Plate

Figure 2-3. Projected areas with oversize beam.

2.3 DEFINITIONS OF BASIC RADIOMETRIC QUANTITIES[1]

Energy is the ability to do work, and most of us have an intuitive understanding of it. Young couples are well aware of the energy of a two-year-old. This is, in fact, kinetic energy, very kinetic. It can be measured by the well-known equation $KE = \frac{1}{2} mv^2$. A relatively famous advertisement for vacuum cleaners uses the concept of potential energy by holding a bowling ball with their vacuum cleaner over the head of the president of the company. Should it let go, he would be impressed with its erstwhile potential energy. In this case, the potential energy is related to the force of gravity and given by the expression $P=mgh$ (mass times the force of gravity times height). Other examples abound: the famous Einstein relation for the energy of mass is well known. Energy has the symbol E, as recommended by the International Standards Organization (ISO), but this is in conflict with both electrical field strength and incident flux density, so Q (quantity of energy) will be used for energy in this text.

---

[1]*ANSI Standard Z7.1-1967(RP-16)*, American National Standards Institute, 1967.

Power is the time rate of change of energy. This is also considered the flux of energy. By convention the symbol P is used for power[1]. However, in radiometry the symbol Φ (for Φlux or phlux or flux) is used for this quantity.

The areal density of power, the flux density, has no generic symbol. However, the power per unit area radiated into the overlying hemisphere is designated the *radiant exitance*. The power received per unit area is called *radiant incidance*. The symbols respectively are M and E. The equations for these are

$$E=\frac{\partial \Phi}{\partial A}, \qquad\qquad M=\frac{\partial \Phi}{\partial A}. \qquad\qquad (2\text{-}3)$$

The two quantities are defined identically; the difference is in the direction of radiation—received or emitted.

The radiant flux per unit solid angle is designated the *radiant intensity* and carries the symbol I. This quantity is usually used with unresolved sources, which are often called *point sources* for short, but it can be more general, designating the flux per unit solid angle from an entire area. The equation is

$$I=\frac{\partial \Phi}{\partial \Omega}, \qquad\qquad (2\text{-}4)$$

where Ω is the symbol for solid angle. Although radiometric terms can be used, misused, and abused in a number of ways, intensity may be the worst. My friend Emil Wolf, and many of his colleagues, use it to mean a flux density in watts per square meter. My fellow Arizonan, Neville Woolf, and his astronomer friends, use it to mean what I call radiance, that is, watts per square meter per steradian. I use it for watts per steradian. Therefore intensity is whatever Wo(o)lf(e) wants it to mean. And it is used in still other senses throughout the technical literature[2].

Finally, the flux per unit projected area and per unit solid angle has been named the *radiance*, and the symbol that is now used is L. The defining equation is

$$L=\frac{\partial \Phi}{\partial(A cos\theta)\,\partial\Omega}. \qquad\qquad (2\text{-}5)$$

---

[2]Palmer, J. M. "Getting Intense on Intensity," Metrologia **30**, 371-372, 1993.

Table 2-1 summarizes these concepts and symbols, and Figure 2-4 illustrates them schematically. For convenience, the table also includes the older symbols [in brackets] that still appear in some useful texts[3].

Table 2-1. Basic radiometric quantities.			
Name	Symbol	Equation	Units
Energy	Q		joule, J
Flux	Φ [P]	$\partial Q/\partial t$	watt, W
Flux Density		$\partial \Phi/\partial A$	
Exitance	M [W]	$\partial \Phi/\partial A$	W m^{-2}
Incidence	E [H]	$\partial \Phi/\partial A$	W m^{-2}
Intensity	I [J]	$\partial \Phi/\partial \Omega$	W sr^{-1}
Radiance	L [N]	$\partial \Phi/\partial A \cos\theta \partial \Omega$	W m^{-2} sr^{-1}

Irradiance

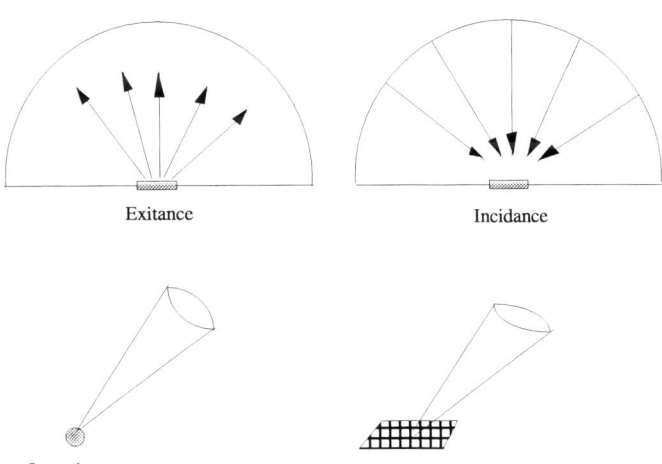

Figure 2-4. Representations of radiometric quantities.

[3]Smith, W. J., *Modern Optical Engineering*, Second Edition, McGraw Hill, 1990.

Some of the symbols may seem a little strange, but they come from the French, who led the naming and development of radiometric standards at the last "go round." The E comes from *eclairage*; the L comes from *luminosité*; the I comes from *intensité*. No one, including my French-speaking friends, can tell me the origin of M.

These are the basic symbols, but other variations arise. These quantities can be distributed spectrally. They can be specific to luminous quantities and they can be photonic quantities. The geometrics are the same for all, as was pointed out by Jones[4] and will be described shortly.

2.4 PHOTONIC RADIOMETRIC QUANTITIES

Photonic quantities start with the photon flux. This is the average time rate of the number of photons. The symbol is subscripted with a q (for quantum) and can be expressed as

$$\Phi_q = \frac{\partial N}{\partial t}, \qquad (2\text{-}6)$$

where N is the number of photons. Rather than watts, the unit for photon flux is number per second and the unit is s^{-1}. The other symbols then are M_q, E_q, L_q, and I_q, with number per second replacing watts in all the rest of the units. For monochromatic radiation each of the energetic quantities can be obtained from the photonic quantities by multiplying by the energy of the photon at the appropriate wavelength.

2.5 SPECTRAL VARIABLES

Spectral quantities represent distributions of the radiometric quantity with respect to a spectral variable. It makes sense to review the different spectral variables before entering a discussion of spectral radiometric quantities.

Perhaps the most familiar spectral variable is wavelength, λ. It is a measure of the minimum distance between two points on a monochromatic wave that have equal phase. The most frequent units are micrometers, μm, and nanometers, nm. The

---

[4]Jones, R. C., "Terminology in photometry and radiometry," Journal of the Optical Society of America **53**, 1314, 1963.

visible part of the spectrum is most often specified in nm, while the infrared is described in μm. The frequency is a representation of the number of cycles per second or waves per unit distance. The frequency, v, in cycles per second is expressed in hertz, Hz; the frequency in waves per unit distance is called the wavenumber, σ, and is usually expressed in waves per centimeter, cm^{-1}. The frequency may also be expressed as an angular, circular, or radian frequency, ω, where ω is $2\pi v$, and the radian wavenumber is $k = 2\pi/\lambda$. Many texts use v with a tilde ~ on top rather than σ, but that introduces considerable typographical difficulties (I could not generate the symbol with WordPerfect). A final frequency that is used is the nondimensional frequency x, which is defined as $c_2/\lambda T$ where c_2 is the second radiation constant and T is the absolute temperature. It is useful and will be used in the theoretical manipulation of blackbody functions.

It should be clear that $\sigma = 1/\lambda$ and $\omega = 1/k$, and this is true if σ is in reciprocal centimeters and λ is in centimeters. However, when the wavelength is in μm, as it frequently is, and the wavenumber is in cm^{-1}, as it most frequently is, the relationship is $\sigma = 10000/\lambda$. Table 2-2 Summarizes these spectral-variable relationships.

Table 2-2. Spectral variables.			
Name	Symbol	Units	Relation
Frequency	v	Hz	
Wavelength	λ	μm	
Wavenumber	σ	cm^{-1}	$\sigma = 10000/\lambda$
Radian wavenumber	k	rad cm^{-1}	$k = 2\pi\sigma$
Radian frequency	ω	rad s^{-1}	$\omega = 2\pi v$
Dimensionless frequency	x		$x = c_2/\lambda T = hc/\lambda kT$
Dimensionless frequency	x		$x = c_2\sigma/kT \cdot hc\sigma/kT$

The proper values for c_2 must be used in the last two relationships, in μm for the wavelength and in cm for the wavenumber.

2.6 SPECTRAL RADIOMETRIC QUANTITIES

These quantities represent distributions with respect to some spectral variable. For purposes of simplification, the spectral variable will be taken as λ, although the discussion applies to any of the other variables discussed above.

Spectral radiance is defined as

$$L_\lambda = \frac{\partial L}{\partial \lambda}. \tag{2-7}$$

It is clearly a distribution of radiance with respect to the wavelength. Mathematically it is a partial derivative. Physically it is the amount of radiance per unit wavelength. It does not have a full physical meaning until either the wavelength variable is specified or until it is multiplied by some spectral bandwidth. It is a very useful and fairly basic function. In fact, the quantity $L_{q\lambda}$ is probably the most fundamental radiometric quantity of all. It is the number of photons per unit time per unit spectral interval, per unit projected area, per unit solid angle. All of the other quantities can be obtained by appropriate integrations.

2.7 LUMINOUS QUANTITIES

Since visible light is so important, a special set of quantities has arisen for use strictly in that domain. These include the lumen, luminous intensity, illuminance, and luminous exitance. It will suffice to indicate the nature of the lumen, as the other geometric relations are analogous to the energetic and photonic quantities. The lumen is defined as

$$\Phi_v = \int V(\lambda)\Phi(\lambda)d\lambda, \tag{2-8}$$

where $V(\lambda)$ is the absolute spectral response of the human eye. This is a "light watt." It is the power that gives rise to a response in the human eye. This is a form of normalization, and normalization will be covered with more generality and detail in Chapter 9.

2.8 FLUOMETRY

In 1963 Jones[5] described the fact that all of these different geometric forms of radiation apply to the different spectral distributions of radiation, the luminous quantities and photonic quantities and more. He called this generalization fluometry (should it have been phluometry, the geometry of Φlux?). He started with a general quantity called Q, that I shall assume for illustration, to represent the number of coeds (Qoeds?). It could be energy, photon number, or something else. Then the *flux* of coeds, Φ, is the time rate of change of their number. Our department chairmen at the Optical Sciences Center have always been interested in this coedic flux, the graduation rate. And then there is both the *incidance* and *exitance* of coeds. Where the flux density is highest, the people watching is the best, and *incidance* describes those arriving, while *exitance* designates those who are going. The *intensity* of coeds would easily be seen to be the rate per solid angle of them leaving each door in Indianapolis after the Arizona championship game at the Final Four in 1997. The *radiance* of coeds (I like that term) is surely the number per unit area and per unit solid angle streaming out of the stadium. The description was in terms of coeds, but describes all students, faculty, and "particles" in general.

If these geometrical characteristics can be described in such a way for coeds—the flux, the incidance, the exitance, the intensity, and the radiance—it can surely be done for photons, for power, for luminous power, for the standard deviation in the rate of photons, for molecules, atoms, and automobiles. This means that the geometric aspects and the spectral aspects of radiometric quantities are separable. Table 2-3 summarizes these concepts in what I hope is a useful way.

The definitions in this table are given in terms of partial derivatives to emphasize the change with respect to a single variable. They can also be written in terms of differentials, e.g.,

$$L_\lambda = \frac{\partial^3 \Phi}{\partial A \cos\theta \, \partial\Omega \, \partial\lambda} = \frac{d^3 \Phi}{dA \cos\theta \, d\Omega \, d\lambda}, \tag{2-9}$$

when there is no ambiguity and when manipulations are to be made.

---

[5]Jones, R. C., "Terminology in photometry and radiometry," Journal of the Optical Society of America **53**, 1314, 1963.

Table 2-3. Summary of radiometric equation definitions.

Type	General	Photons	Energy	Spectral	Visible
Quantity	Q	N	Q	Q_λ	
Flux	Φ	$\Phi_q = \partial N/\partial t$	$\Phi_u = \partial Q/\partial t$	$\Phi_\lambda = \partial Q/\partial\partial\lambda$	$\Phi_v = \partial Q_v/\partial t$
Flux Density					
Incidance	$E = \partial\Phi/dA$	$E_q = \partial N/(\partial t\,\partial A)$	$E_u = \partial\Phi/\partial A$	$E_\lambda = \partial^2\Phi/(\partial A\partial\lambda)$	
Exitance	$M = \partial\Phi/\partial A$	$M_q = \partial^2 N/(\partial t\partial A)$	$M_u = \partial\Phi/\partial A$	$M_\lambda = \partial^2\Phi/(\partial A\partial\lambda)$	$M_\lambda = \partial^2\Phi/(\partial A\partial\lambda)$
Intensity	$I = \partial\Phi/\partial\Omega$	$I_q = \partial^2 N/(\partial t\partial\Omega)$	$I_u = \partial\Phi/\partial\Omega$	$I_\lambda = \partial^2\Phi/(\partial\Omega\,\partial\lambda)$	$I_\lambda = \partial^2\Phi/(\partial\Omega\,\partial\lambda)$
Radiance	$L = \partial^2\Phi/(\cos\theta\partial A\partial\Omega)$	$L_q = \partial^3 N/(\partial t\cos\theta\,\partial A\partial\Omega)$	$L_u = \partial^2\Phi/\cos\theta\,\partial A\partial\Omega$	$L_\lambda = \partial^3\Phi/(\cos\theta\partial A\partial\Omega\,\partial\lambda)$	$L_\lambda = \partial^3\Phi/(\cos\theta\partial A\partial\Omega\,\partial\lambda)$

IRRADIANCE

2.9 The Chinese Restaurant System[6]

I don't know if this actually happened on a napkin, but Jon Geist and Ed Zaleski concocted this scheme of nomenclature to be as precise as they knew how in describing these various radiometric quantities. They decided that there were four aspects to each. There was the process, the spectral aspect, the quantity, and the fluometry. It is best described by a set of examples. These are shown in Table 2-4. The first row indicates the type of descriptor; the next four are examples.

Table 2-4. The Chinese Restaurant Nomenclature System			
Process	**Spectrum**	**Quantity**	**Flux Geometry**
Scattered	Spectral	Energy	Radiance
Reflected	Bandlimited	Photon	Irradiance
Transmitted	Erythemal		Fluence
Emitted	Luminous		Exitance

This illustrates that the process, spectrum, quantity, and flux geometry are all separable quantities. (Erythemal is the part of the spectrum that is effective in producing sunburn—a sunburn watt.) The first two need the specification energy or photon, but the second two have the quantity defined. The Chinese Restaurant System is probably best used in the defining-relationship part of any discussion. It is surely cumbersome and should be replaced subsequently in any discussion or publication with a more shorthand notation.

2.10 Recap

There are four geometric terms in radiometry: *radiance*, L, the flux per unit projected area and solid angle; *radiant exitance*, M, the flux per unit area emitted into the overlying hemisphere; E, the flux per unit area received from the overlying hemisphere; and *intensity*, I, the flux per unit solid angle. These quantities can relate to the flux of energy, the flux of photons, or the flux of visible light, indicated by the subscripts u, q, and v (although often there is no subscript used with energetic quantities). They can also relate to the total spectrum, the

---

[6]Geist, J. and E. Zalewski, "Chinese restaurant nomenclature for radiometry," Applied Optics **12**, 435, 1973.

spectrum integrated over some spectral band, or essentially monochromatic radiation. These are indicated, where necessary, by no subscript for the full spectrum, $\Delta\lambda$, and λ, respectively.

One should read and write about these with great care. The use of the Chinese Restaurant System is a good way to make sure that radiometric quantities are well defined.

CHAPTER 3

RADIATIVE TRANSFER

This chapter explains some of the basics of radiative transfer, and it provides explanations of some of the concepts that are related to the transfer of radiation among and between surfaces and optical systems.

3.1 THE FUNDAMENTAL EQUATION OF RADIATIVE TRANSFER

The transfer of radiation in a vacuum environment from one surface to another, no matter how complicated the two surfaces, can be described succinctly by the differential form of the fundamental equation of radiative transfer. It is

$$d\Phi = L \frac{dA_1 \cos\theta_1 \, dA_2 \cos\theta_2}{\rho^2} \, ,$$

(3-1)

where L can be the radiance from either surface to the other, or the net radiance, A is an area, θ is the angle between a surface normal and the line of sight between the surfaces, ρ is the line-of-sight distance, and the subscripts indicate the first and second surfaces. This is illustrated in Figure 3-1.

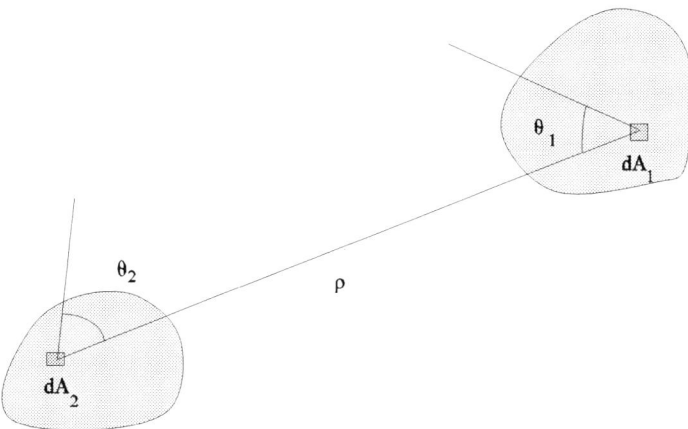

Figure 3-1. The fundamental equation of radiative transfer.

The radiance is not of differential form, because by definition the radiance is a distribution and is of itself differential in nature. A useful exercise is to substitute the expressions for radiance and for solid angle back into this expression to obtain the tautology:

$$d\Phi = L\frac{dA_1\cos\theta_1 dA_2\cos\theta_2}{\rho^2} = \frac{d\Phi}{dA_1\cos\theta_1 d\Omega_{21}}\cos\theta_1 dA_1 d\Omega_{21} = d\Phi \ . \qquad (3\text{-}2)$$

This, of course, is why the radiance was defined as it was. There are other forms of this equation that are used and useful. By inserting the definition of one solid angle,

$$d\Omega_{12} = dA_1\cos\theta_1/\rho^2 \ , \qquad (3\text{-}3)$$

where the subscripts 1 and 2 mean the solid angle of area 1 as seen from area 2, a new form is found:

$$d\Phi = LdA_1\cos\theta_1 d\Omega_{21} \ . \qquad (3\text{-}4)$$

Similarly, one can substitute the expression for the other solid angle,

$$d\Omega_{21} = dA_2\cos\theta_2/\rho^2 \ , \qquad (3\text{-}5)$$

and the next form becomes

$$d\Phi = LdA_2\cos\theta_2 d\Omega_{12} \ . \qquad (3\text{-}6)$$

The projected solid angle, the projected area times the "other" cosine, can also be introduced, to obtain

$$d\Phi = LdA_1 d\Omega_{p21} = LdA_2 d\Omega_{p12} \ . \qquad (3\text{-}7)$$

The final, useful form is

$$d\Phi = LdZ, \qquad\qquad (3\text{-}8)$$

where dZ is the differential throughput. It is clearly the multiplier of L in Eq. (3-1). It is also called the *etendué*, and the *area-solid angle* or *A-Ω product*. (Of course, this is the projected area—solid-angle product or the area—projected-solid-angle product). The word *etendué*, from the French, means extent or size, while the others come from the usual symbols and the words for the integrated form of the throughput. The Germans use *Lichtleitwert*, or light transference value. Others like throughputability, transferent, extendent, and more. There is no well-accepted symbol or name for it[1]; I like Z for zap or zip or zing or zest, even though Steele[1] likes G.

Equation (3-8) is a very nice form. It separates the expression for transfer into two terms, the radiance and the throughput. The radiance is a property of the source and the throughput is a property of the geometry, usually describing the nature of the instrumental receiver. Each is invariant throughout the optical system, as will be shown in Chapter 5.

3.2 LAMBERTIAN EMITTERS

The fundamental equation can be applied to a very useful fiction, the lambertian radiator. This radiator has the property that its radiance is independent of the angle into which the radiation is directed, i.e., isotropic, so that $L \neq L(\theta)$. This example is a calculation of the radiant exitance from an isotropic radiator. Figure 3-2 shows the geometry. A small differential area dA radiates with radiance L into the overlying hemisphere that has a radius R. Therefore the fundamental equation becomes

$$d\Phi = L\frac{dA_1\cos\theta_1 dA_2\cos\theta_2}{\rho^2} = LdA\sin\theta d\theta d\varphi\cos\theta. \qquad (3\text{-}9)$$

The expression on the right arises from the fact that the differential element of

---

[1]Steele, W. H., "Luminosity, Throughput or Etendue?" Further Comments, Applied Optics **14**, 252, 1975.

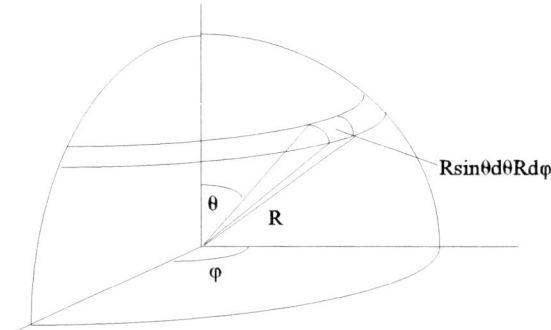

Figure 3-2. The geometry of a lambertian emitter and the overlying hemisphere.

area on the sphere is $R\sin\theta Rd\varphi$, and because the (constant) radius, R, is perpendicular to that element, the cosine is 1. This step immediately shows that the differential solid angle is

$$d\Omega = \sin\theta d\theta d\varphi, \qquad (3\text{-}10)$$

and the projected solid angle is

$$d\Omega_p = \sin\theta d\theta d\varphi \cos\theta . \qquad (3\text{-}11)$$

So the emitted flux density M is the power per unit area dA and is therefore the integral of the radiance times the projected solid angle over the hemisphere:

$$M = \frac{\Phi}{dA} = L\int_0^{2\pi}\int_0^{\pi/2}\sin\theta\cos\theta d\theta d\varphi . \qquad (3\text{-}12)$$

Recall that in this example the radiance is independent of angle, by assumption. Integration around the sphere in an annulus gives

$$M = \frac{\Phi}{dA} = L\int_0^{\pi/2}\int_0^{2\pi}\sin\theta\cos\theta d\theta d\varphi. \qquad (3\text{-}13)$$

Then,

$$M = 2\pi L \int_0^{\pi/2} \sin\theta\cos\theta \, d\theta = 2\pi L \left[\frac{1}{2}\sin^2\theta\right]_0^{\pi/2} = \pi L \left[\sin^2\theta\right]_0^{\pi/2} = \pi L. \qquad (3\text{-}14)$$

The radiant exitance is π times the radiance, or in the way it is usually said, the radiance is the radiant exitance divided by π. This is an interesting result that is often gotten wrong and often not understood. The (erroneous) result is often obtained by an argument that says there are 2π steradians in a hemisphere, and since the radiance is uniform and a flux density per steradian, there must be an exitance of $2\pi L$. Physically, the interaction is between an element on the overlying sphere and the *projected* area of the emitter. On average, the projected area is one-half the physical area. So the relationship is only π. Another way to look at it is that there are π steradians in a *projected* hemisphere. The integration can also be done for a conical beam that is less than an entire hemisphere. This result is useful in some applications:

$$M = \pi L \left[\sin^2\theta\right]_0^{\Theta} = \pi L \sin^2\Theta. \qquad (3\text{-}15)$$

Although there really is no such thing as a lambertian source (free lunch or an uncluttered horizontal surface), it is often a useful approximation.

By reasonable extension, the surface dA could be a lambertian reflector or a lambertian receiver.

3.3 TRANSFER BETWEEN A DIFFERENTIAL ELEMENT AND A DISK

The second example is also a "twofer." Both the process and the result are useful. Figure 3-3 shows the geometry of a differential element and a disk. In this case, rather than calculate the relationship between a radiance and an exitance or incidence, we calculate the throughput of the system. The process is to start with the fundamental equation of transfer,

$$d\Phi = L \frac{dA_1 \cos\theta_1 \, dA_2 \cos\theta_2}{\rho^2} = L \, dZ. \qquad (3\text{-}16)$$

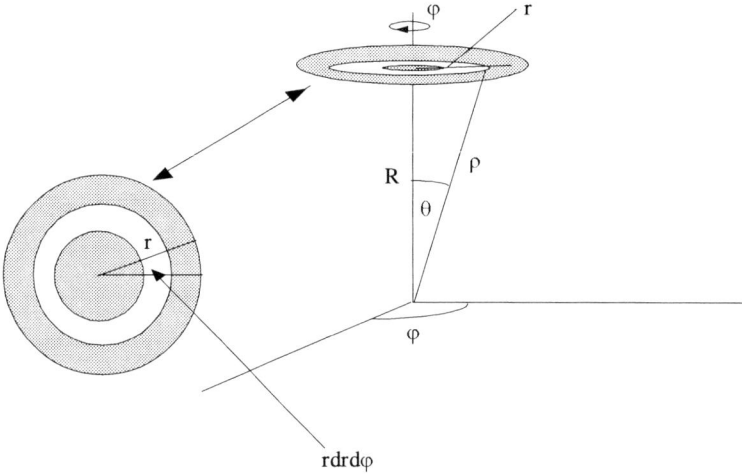

Figure 3-3. Geometry for the lambertian disk.

Then the differential throughput is given by

$$dZ = \frac{dA_1 \cos\theta_1 \, dA_2 \cos\theta_2}{\rho^2} .$$

(3-17)

The disk is an axial distance R away from the differential element, and the angle the line of sight makes from the axis is θ. The proper terms based on the geometry need to be inserted. This is the process for solving such problems. Assume that dA_1 is the differential area dA, and then write the differential element on the disk as $r \, d\varphi \, dr$, where r and φ are the polar coordinates on the disk. The line-of-sight-distance varies; it is not a constant R as in the case of the sphere, but it is R divided by $\tan\theta$. The radius r on the disk is related to the axial distance R by $R\tan\theta$. Therefore,

$$dZ = \frac{dA \cos^2\theta \, r \, dr \, d\theta \, d\varphi}{R^2 \sec^2\theta} = \frac{dA \cos^2\theta \, R\tan\theta \, R\sec^2\theta \, d\theta \, d\varphi}{R^2 \sec^2\theta} .$$

(3-18)

Almost all of the trig functions in the numerator and the denominator cancel, so that

$$dZ = dA\sin\theta\cos\theta d\theta d\varphi = dA d\Omega' .$$

(3-19)

This is exactly the same differential projected solid angle as we calculated for the sphere. This is a nice confirmation that the value of a solid angle is dependent on the periphery and not on the topography—the disk and the sphere have the same periphery but different topography.

But there is more. The irradiance can be written as the radiance times the throughput divided by the area (after a little integration):

$$E = \frac{LA}{R^2}\sin^2\theta = \frac{LA}{R^2}\frac{r^2}{r^2+R^2} = \frac{I}{R^2}\frac{1}{1+(r/R)^2} = \frac{I}{R^2}\left[1-(r/R)^2+(r/R)^4-+\cdots\right].$$

(3-20)

This interesting form shows that the well-known inverse square law is only an approximation. The real inverse square law was *enacted* in terms of point sources, sources with infinitesimal areas. It applies as long as the ratio of the radius of the source is small enough with respect to the distance.

3.4 OTHER GEOMETRIES AND DISTRIBUTIONS

The two examples of radiation calculation have both been with lambertian and circularly symmetric sources. In this section we show how these two assumptions can be relaxed, that is, how to calculate radiative transfer for more general radiances and geometries.

The radiance from a body is, in general, a function of both the azimuth and the elevation angles, θ and φ. A lambertian emitter is one in which the radiance is independent of angle, constant with angle, the same in all directions. Thus, a lambertian radiator has a radiance L, $(\neq L(\theta,\varphi))$.

The radiance in an idealized collimated beam has a value L in the beam and zero elsewhere. Thus it can be represented in one dimension by either $L(\theta_0)$ or $L\delta(\theta-\theta_0)$, where $\delta(x)$ is the delta function. In practice, of course, this idealization is never achieved and the radiance in a collimated beam would be infinity.

The radiance, $L(\theta,\varphi)$, is a general function of the angle. It is convenient to represent this is in terms of a power series of cosines. For one dimension, and assuming circular symmetry so that $L(\theta,\varphi) = L(\theta)$,

$$L(\theta) = \sum_{n=0}^{N} L_n \cos^n \theta. \tag{3-21}$$

The coefficients L_n correspond to the different orders in the power series. The lambertian surface is represented by the constant term for which $n=0$. Figure 3-4 shows the distributions for several of these radiators.

Then the relationship between radiant exitance and radiance is

$$E = 2\pi \int \sum_{n=0}^{N} L_n \cos^{n+1}\theta \sin\theta d\theta = -2\pi \sum_{n=0}^{N} L_n \int \cos^{n+1}\theta d\cos\theta = -\pi \sum L_n [\cos^{n+2}]_0^\Theta. \tag{3-22}$$

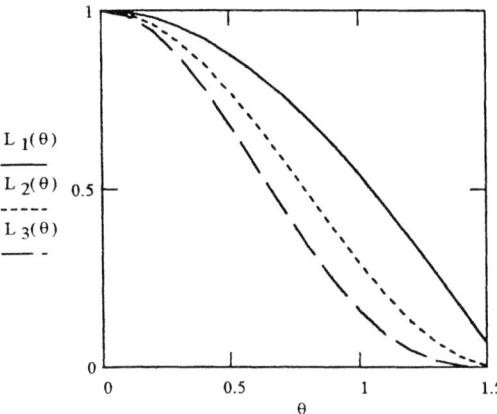

Figure 3-4. Several angular distributions. Lambertian is a horizontal line; then orders 1, 2, and 3 are increasingly decreasing with angle.

When Θ is $\pi/2$, representing the full hemisphere, the result is

$$E = \pi \sum L_n 1^{n+2} = \pi \sum L_n \tag{3-23}$$

and when n is 0, the result is the same as for the lambertian source, as it should be. This representation can be used for lambertian emitters, reflectors, and transmitters. Many texts describe a lambertian radiator or receiver as a $\cos\theta$ radiator or receiver. In so doing, they are describing the angular dependence of the flux density, not that of the radiance.

The world is full of different geometric shapes, and consideration of all of them could take at least all of the remaining space of this text. Consideration of a rectangle is in order, and delegation of other configurations to an appendix is appropriate. Consider the radiation from a differential element elevated above the corner of a rectangle and parallel to the rectangle, as shown in Figure 3-5.

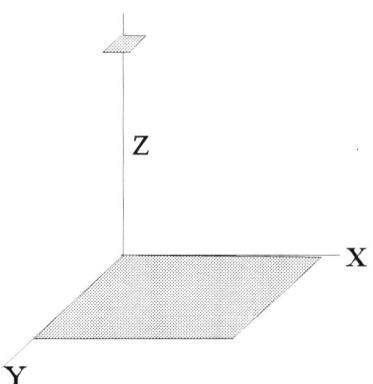

Figure 3-5. A differential element and rectangle parallel to it.

The fundamental equation of radiative transfer is

$$dP = L \frac{dA_1 \cos\theta_1 dA_2 \cos\theta_2}{\rho^2}.$$

(3-24)

The irradiance on dA_1 can be calculated by dividing through by dA_1 and the two cosines are equal by virtue of alternate interior angles, so

$$dE = L \frac{dA_2 \cos^2\theta}{\rho^2}.$$

(3-25)

The differential area on the rectangle is $dxdy$ and the variable distance ρ is given by the three-dimensional pythagorean theorem, so that

$$dE = L \frac{\cos^2\theta dxdy}{x^2 + y^2 + Z^2} = L \frac{Z^2 dxdy}{\rho^4},$$

(3-26)

where Z is the fixed altitude above the rectangle, and the cosine is equal to Z/ρ. Finally,

$$E = L \int_0^X \int_0^Y \frac{\cos^2\theta}{x^2+y^2+Z^2} dxdy = L \int_0^X \int_0^Y \frac{Z^2}{\rho^4} dxdy. \qquad (3\text{-}27)$$

There is no azimuthal angle, φ, in this formulation.

These two analyses can be combined, that is, we can obtain the proper expression for the irradiance at a differential area a distance Z above the rectangle for a non-lambertian radiator:

$$E = \sum L_n \int_0^X \int_0^Y \frac{Z^{n+2}}{\rho^{n+4}} dxdy, \qquad (3\text{-}28)$$

since $\cos^n\theta$ is Z/ρ.

This can be extended to two rectangles and other geometries, as is done in the Appendix.

3.5 RECAP

The fundamental equation of radiative transfer can be written as the radiance L times the throughput Z, where Z is given by

$$Z = \frac{n^2 dA_1 \cos\theta_1 dA_2 \cos\theta_2}{\rho^2}. \qquad (3\text{-}29)$$

Both L/n^2 and Z are invariant throughout the optical system. The radiance can be from either surface to the other or the net radiance between them. A simple way to remember this important relationship is that it is the product of the projected areas over the square of the distance between them. Several examples of the use of this equation to evaluate the transfer between certain geometries have been given, and two principles have been developed. One is that the radiance is the radiant exitance divided by π for a lambertian surface. This applies to emission and to reflection. The other is that the inverse square law only applies when the extent of the surface is much smaller than the distance. Finally, there is no such thing as a lambertian surface, but real surfaces can be represented by a power series of cosines. In combination these techniques allow the calculation of the transfer between any reasonable surfaces, and the unreasonable ones can be attacked by a finite-element process.

CHAPTER 4

TRANSMISSION, REFLECTION, EMISSION,

AND ABSORPTION

A material may transmit, reflect, emit or absorb radiation, and generally does more than one of these at a time. These properties have spectral and geometric characteristics that are described by appropriate adjectives. The properties come in pairs. Transmission and reflection differ only in directionality. In a very real sense they are only 180 degrees apart. By Kirchhoff's law it can be shown that, under identical conditions, the emission of a body is equal to its absorption.

4.1 SOME DEFINITIONS

This is the section on *ivities, ances,* and *ions.* It is generally accepted that the ending *ion* signifies a process. Emission is the process of radiating. Transmission is the process of transmitting. The same with absorption and reflection. An *ance* ending is said to indicate the property of a particular sample. Thus, one reports the reflectance of sample 2306 as measured on April Fool's Day. The *ivity* ending is meant to signify the property of the "generic brand," an idealized sample that represents all such samples of that material. There has been controversy about whether there is a meaningful difference between a reflectance and a reflectivity[1]. For reflectivity, the sample must be pure, clean, and smooth enough that the reflection is not affected by any lack thereof. But a body always has some roughness, and the effects of this roughness always can be measured. In this text the *ance* and *ivity* forms of these words will be used interchangeably, but the reader should understand that some workers make a distinction (or distinctance or even distinctivity!).

4.2 THE CONSERVATION OF POWER

Figure 4-1 shows light incident on a plane parallel plate. Some of it is reflected, some is absorbed, and some is transmitted. The sum of these powers must add up to the incident power if power conservation is to be maintained. Then

---

[1]Richmond, J. C., "Rationale for emittance and reflectivity," Applied Optics **21**, 1, 1982; W. L. Wolfe, "Proclivity for emissivity," Applied Optics **21**, 1, 1982.

$$\Phi_0 = \Phi_a + \Phi_r + \Phi_t .$$ (4-1)

Division yields

$$1 = \frac{\Phi_a}{\Phi_0} + \frac{\Phi_r}{\Phi_0} + \frac{\Phi_t}{\Phi_0} = \alpha + \rho + \tau ,$$ (4-2)

where the absorptivity, α, reflectivity, ρ, and transmissivity, τ are defined by this equation. Several variations are available with this thought experiment. Imagine that the radiation is almost monochromatic at wavelength λ. Then, as long as equilibrium is maintained (there is no frequency doubling, no fluorescence, etc.), then conservation of energy requires that

$$\frac{\Phi_a(\lambda)}{\Phi_0(\lambda)} + \frac{\Phi_r(\lambda)}{\Phi_0(\lambda)} + \frac{\Phi_t(\lambda)}{\Phi_0(\lambda)} = \alpha(\lambda) + \rho(\lambda) + \tau(\lambda) = 1 .$$ (4-3)

A similar argument applies to each of the components of linear or circular polarization, as long as the polarization is conserved. This may not be true with metallic plates that can introduce elliptical polarization, and it is surely not true with optically active materials.

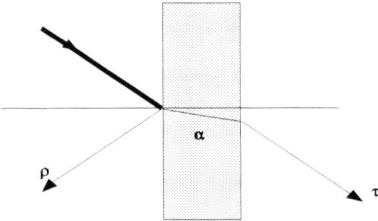

Figure 4-1. Transmittance, τ, absorptance, α, and reflectance, ρ.

4.3 KIRCHHOFF'S LAW[2]

This law in essence equates absorptivity and emissivity. It will make the description of these, given below, half as long, and it has other very important uses.

Consider the cavity shown in Figure 4-2, and assume that it is in thermal equilibrium, all of it at the same temperature, with completely opaque sides.

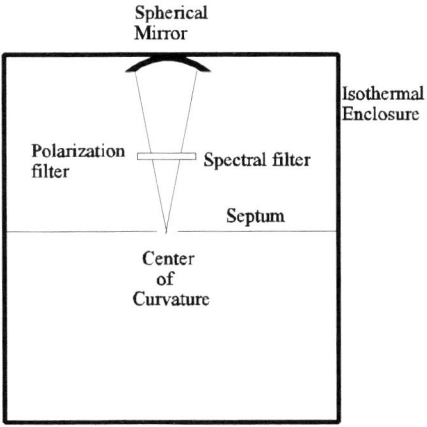

Figure 4-2. The Kirchhoff box.

Then the radiation that goes from the bottom to the top must be exactly equal to the radiation that goes from the top to the bottom. Otherwise the equilibrium would be violated. Start by placing on the top an opaque mirror with a center of curvature at the hole in the septum. Equilibrium is maintained if it is perfectly reflecting, since all the radiation that goes to it is reflected back. But if it is not perfectly reflecting, any deficiency in its reflectivity must be made up exactly by emissivity. Thus using

$$\varepsilon = 1 - \rho, \quad \alpha + \rho + \tau = 1, \quad \tau = 0, \tag{4-4}$$

[2]Kirchhoff, G., "On the relation between the radiating and absorbing powers of different bodies for light and heat," Philosophical Magazine, S4, **20**, 1, 1860, translated from Poggendorf Annalen CIX, 275.

there results Kirchhoff's law, which is

$$\alpha = \varepsilon .$$ (4-5)

Thus far we have dealt with total quantities, based on the conservation of energy. It is not hard to see that with both the plate and with the enclosure, a spectral filter can be used to restrict the entire upward power to a narrow band, so that

$$\alpha(\lambda) = \varepsilon(\lambda) .$$ (4-6)

A similar argument applies with a polarizer, and one can even do it directionally with certain limitations. Thus, one can state Kirchhoff's law very succinctly as absorptivity equals emissivity. More carefully, one must state that under identical conditions the spectral emissivity equals the spectral absorptivity, the total emissivity equals the total absorptivity, and each polarization component of either spectral or total emissivity equals that of the equivalent absorptivity. The identical conditions requirement includes, among other things, that the emissivity and absorptivity are measured for a sample at the same temperature and in the same spectral band. These conditions are described elsewhere[3].

4.4 ABSORPTION AND EMISSION

The absorption and emission properties of a material are described by its emissivity and absorptivity. Only emissivity will be discussed here. Absorptivity follows the same arguments by Kirchhoff's law.

Hemispherical spectral emissivity is defined as the ratio of the radiant exitance of a real body to that of an ideal body, often called a blackbody because an ideal radiator has an emissivity of 1 and its absorptivity is therefore also 1; a perfect radiator is a perfect absorber. The expressions are

$$\varepsilon_h(\lambda) = \frac{M(\lambda)}{M^{BB}(\lambda)}, \quad \varepsilon_h(\sigma) = \frac{M(\sigma)}{M^{BB}(\sigma)},$$ (4-7)

where M is the radiant exitance of a real body, M^{BB} is the radiant exitance of a blackbody, λ is wavelength, and σ is the wave number. This is the ratio of the

---

[3]Palmer, J. M., Chapter 25 in M. E. Bass, E. W. Van Stryland, D. R. Williams, and W. L. Wolfe, Eds., *Handbook of Optics*, McGraw Hill, 1995.

power per unit area per spectral interval into an overlying hemisphere to the same quantity from an ideal radiator, a blackbody.

The directional, spectral emissivity is defined as

$$\varepsilon_d(\lambda,\theta,\varphi)=\frac{L(\lambda,\theta,\varphi)}{L^{BB}(\lambda)}, \quad \varepsilon_d(\sigma,\theta,\varphi)=\frac{L(\sigma),\theta,\varphi}{L^{BB}(\sigma)} \quad . \tag{4-8}$$

There are two other types of spectral emissivities that can be considered. One is the spectral, weighted, average emissivity:

$$\bar{\varepsilon}_d=\frac{\int\varepsilon(\lambda)L^{BB}(\lambda,T)d\lambda}{\int L^{BB}(\lambda,T)d\lambda} \quad . \tag{4-9}$$

The other is the total emissivity:

$$\bar{\varepsilon}_d=\frac{\int_0^\infty\varepsilon(\lambda)L^{BB}(\lambda,T)d\lambda}{\int_0^\infty L^{BB}(\lambda,T)d\lambda} \quad . \tag{4-10}$$

Both of these are weighted by the ideal-radiator spectral distribution, which is a function of temperature. The integrals could equally well be over the wavenumber. The spectral emissivity distributed in wavelength is not the same as the spectral emissivity as a function of wavenumber, but the weighted averages are identical. An example of how the spectral distribution of the source affects the average emissivity is shown in Figure 4-3. It was generated by assuming a detector that has a linearly increasing responsivity with wavelength measures blackbody radiation of the indicated temperatures. Such might be the case with a mercury cadmium telluride detector measuring a blackbody from room temperature to the maximum operating temperature of tungsten. The variation is larger at lower temperatures.

In summary, there are two different types of emissivity (and absorptivity) based on geometry, the hemispherical and the directional. There are also three different emissivities based on the spectrum: spectral, total, and weighted average. Thus, there are six different kinds of emissivity (and absorptivity).

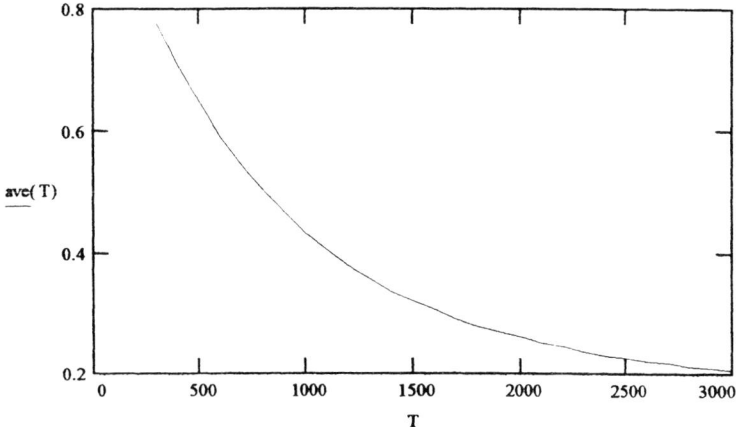

Figure 4-3. The effect of spectral distribution on average emissivity.

4.5 TRANSMISSION AND REFLECTION

There are three different spectral types of transmission and reflection as well. There are a number of different geometries[4].

The most familiar reflectivity is undoubtedly specular reflectivity, illustrated in Figure 4-4 . It is just the ratio of the reflected power to the incident power, but the power must be that which is in the beam. Thus,

$$\rho_s(\theta_i, \varphi_i) = \frac{\Phi^{ref}}{\Phi^{inc}} \ . \qquad (4\text{-}11)$$

The specular reflectivity is a function of only the two angles of incidence. The angles of reflection are, of course, equal to those of incidence by Snell's law. The figure need not be three dimensional since part of the law of specular reflection is that the reflected beam is coplanar with the incident beam. The angle φ is the angle between the plane of incidence and a reference on a surface that is not homogeneous or isotropic.

---

[4]Nicodemus, F. E., J. C. Richmond, J. J. Hsia, I. W. Ginsberg, and T. Limperis, *Geometrical considerations and nomenclature in reflectance,* National Bureau of Standards, U. S. Department of Commerce, 1977; Judd, D. B., "Terms, Definitions and Symbols in Reflectometry," Journal of the Optical Society of America **57,** 445, 1967.

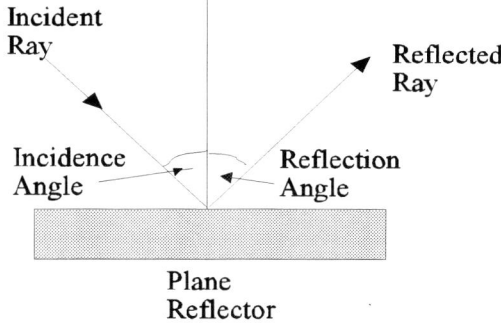

Figure 4-4. Specular reflection.

A second type of reflectivity is the directional-hemispherical reflectivity, sometimes called just the hemispherical reflectivity. It is shown in Figure 4-5 and is defined as

$$\rho_{dh} = \frac{M^{ref}}{\Phi^{inc}/A}.$$ (4-12)

Usually the incident light is a beam and is incident from a particular direction. The reflected light is that which is collected over the entire overlying hemisphere. The hemispherical-directional reflectivity is just the inverse of directional-hemispherical, both physically and mathematically.

The bidirectional reflectivity is a different sort of thing. It is defined as the reflected radiance divided by the incident irradiance[5],

$$\rho_{bd}(\theta_i,\varphi_i;\theta_r,\varphi_r) = \frac{L^{ref}(\theta_r,\varphi_r)}{E(\theta_i,\varphi_i)}.$$ (4-13)

[5]Nicodemus, F. E., "Directional reflectance and emissivity of an opaque surface," Applied Optics **4**, 767, 1965.

It has units of reciprocal steradians and does not have a maximum value of one, as do all the previous reflectivities. It is illustrated in Figure 4-6. Another way to think about this bidirectional reflectivity is that it is "normal" reflectivity per unit solid angle, i.e.,

$$\rho_{bd}(\theta_i, \varphi_i; \theta_r, \varphi_r) = \frac{M^{ref}(\theta_r, \varphi_r)}{E^{inc}(\theta_i, \varphi_i)\Omega} = \frac{\rho}{\Omega}. \tag{4-14}$$

The maximum value can be found by a *gedanken* experiment, a thought experiment. Imagine a perfectly smooth surface with 100% specular reflectivity and an incident, perfectly collimated beam. (Both of these are pure fiction.) Then the specular reflectivity M/E is 1 and the solid angle is 0. The maximum value of the bidirectional reflectivity of this hypothetical surface is infinity! In practice very high values, approaching a million, are observed with good mirrors.

There is a nice relationship between directional-hemispheric reflectivity and bidirectional reflectivity for a lambertian (isotropic) reflector. It is

$$\rho_{bd} = \frac{\rho_{dh}}{\pi}. \tag{4-15}$$

The radiation from a lambertian reflector is distributed uniformly over the hemisphere, and the value of the projected hemisphere is π. Integration will also do it:

$$\rho_{dh} = \frac{M}{E} = \frac{\int_{hemi} L d\Omega}{E} = \frac{L\int_0^{2\pi}\int_0^{\pi/2}\sin\theta\cos\theta d\theta d\varphi}{E} = \frac{\pi L}{E}. \tag{4-16}$$

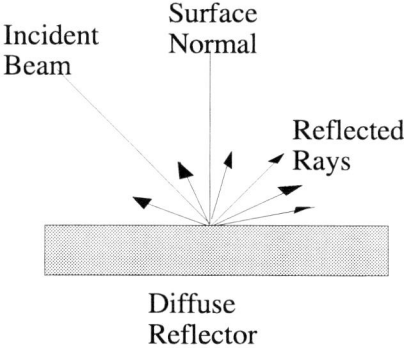

Figure 4-5. Hemispherical reflectivity.

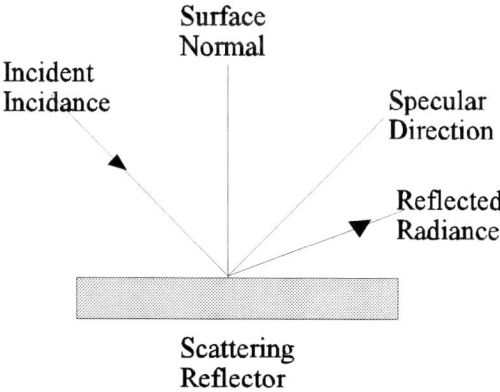

Figure 4-6. Bidirectional reflectance.

4.6 SOME EXAMPLES

It may be useful to have a few representative types of reflectors in mind. A mirror is a specular reflector with high reflectivity, while a piece of white paper is a diffuse (almost lambertian) reflector with high reflectivity. Black velvet is a diffuse reflector of low reflectivity, a diffuse black, while black glass is a good example of a specular reflector of low reflectivity, a specular black. Of course, materials in general range the gamut of almost perfectly specular to almost perfectly diffuse and are all shades of gray.

Another interesting calculation is a less idealized relationship between directional hemispherical reflectivity and bidirectional reflectivity. Assume a monochromatic beam at 10 μm that has been collimated to within the diffraction limit by a circular mirror of 10 cm diameter. Then the bidirectional reflectivity is the hemispherical reflectivity ρ_h divided by the solid angle, which is

$$\Omega = \frac{\pi}{4}\left(\frac{2.44\lambda}{D}\right)^2 = \frac{\pi}{4}\left(\frac{2.44\times10^{-6}}{0.1}\right) = 4.675\times10^{-10}. \qquad (4\text{-}17)$$

Then the bidirectional reflectivity is $2.14\times10^9\rho_h$. For a ρ_h of 1 this is a reciprocal gigasteradian.

4.7 RELATIONSHIPS AMONG TRANSMISSIVITY, REFLECTIVITY, ABSORPTIVITY, AND EMISSIVITY

It has already been shown that transmissivity and reflectivity are essentially the same, and so are absorptivity and emissivity. The question now is how does reflectivity relate to emissivity and absorptivity, and similarly for transmissivity and these two quantities. It should be apparent that it is enough to show the relationship for one pair.

There must be restrictions placed on reflectivity if it is to relate directly to emissivity, because one depends on two angles and is bidirectional, while the other is unidirectional. Based on this simple argument, it is apparent that $\varepsilon = 1-\rho_{dh}$. The emissivity is the one's complement of the directional-hemispherical reflectivity. One of the two angles was eliminated by insisting on the hemispherical geometry.

4.8 RECAP

Radiometric properties of materials include transmission, reflection, absorption, and emission. The *ion* ending indicates a process; the ratios represented by these quantities have endings of *ance* and *ivity*. The properties divide nicely into transmission and reflection, which are different by geometry only, and absorption and emission, which are alike based on Kirchhoff's law. They can each be described as spectral, bandlimited, and full spectrum, and as hemispherical and directional. Absorption and emission are described unidirectionally, the direction to which or from which they emit or absorb. Reflection and transmission need both the incident and the reflected or transmitted directions in their description; they are bidirectional quantities.

CHAPTER 5

RADIANCE [1]

Since radiance is the fundamental radiometric quantity, this chapter is devoted to some of its properties. Radiance is invariant through an optical system, but when there is a change in the medium the radiance divided by the square of the refractive index is invariant (reduced radiance). All the other radiometric terms can be obtained from radiance by appropriate integrations. Not only is reduced radiance invariant, but the throughput is also invariant. Both of these invariance assertions will be proven in more than one way.

5.1 THE INVARIANCE OF RADIANCE IN A VACUUM

Consider the transfer of radiation from a small source of area A at the left in Figure 5-1.

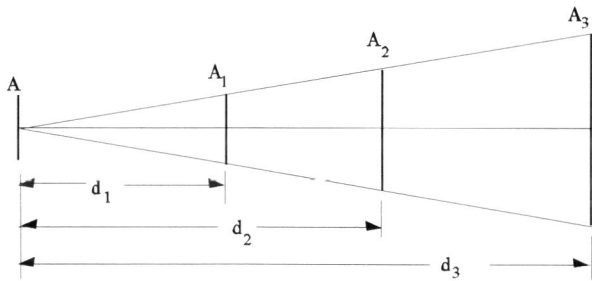

Figure 5-1. Conservation of radiance in a vacuum.

The distances of the several areas are d_1, d_2, etc. By the first law of thermodynamics, the power must be the same on area A_2 as it is on A_1 as it is on A_3 as it is on all of the areas. Only the flux *densities* change. Call this invariant

[1]Nicodemus, F. E., "Radiance," American Journal of Physics **31**, 368, 1963.

power a flux Φ. Then the flux on A_1 is $L\,A\,A_1/d_1{}^2$. The same calculation applies to the other areas, e.g., the flux on A_2 is $L\,A\,A_2/d_2{}^2$. These powers must all be the same (in a nonabsorbing, nonscattering medium). But, by the geometry of similar triangles,

$$\frac{A_1}{d_1^2}=\frac{A_2}{d_2^2}=\cdots. \tag{5-1}$$

Then

$$\Phi=L_1\frac{AA_1}{d_1^2}=L_2\frac{AA_2}{d_2^2}=L_3\frac{AA_3}{d_3^2}=\cdots. \tag{5-2}$$

Now $A{=}A$ and $A_1/d_1^2 = A_2/d_2^2 = A_3/d_3^2 = ...$Therefore $L_1{=}L_2{=}L_3$ ...; the radiance must be constant. If the surfaces were tilted, cosines would enter, but the result would be the same.

5.2 INVARIANCE OF (REDUCED) RADIANCE ACROSS AN INTERFACE[2]

Figure 5-2 shows the geometry of a pencil of light that strikes a plane surface of a plate that has a different refractive index. It will be assumed that the first medium is a vacuum. Snell's law is

$$\sin\theta_1 = n\sin\theta_2 . \tag{5-3}$$

Differentiation yields

$$\cos\theta_1 d\theta_1 = n\cos\theta_2 d\theta_2 . \tag{5-4}$$

Multiplication of these together and by $d\varphi$ then yields

---

[2]Liebes, S., Jr., "On the ray invariance of B/n²," *American Journal of Physics* **37**, 932, 1969; A. Arkangy, "Liouville's theorem and the intensity of beams," ibid, **25**, 519, 1957.

$$\sin\theta_1\cos\theta_1 d\theta_1 d\varphi = n^2 \sin\theta_2 \cos\theta_2 d\theta_2 d\varphi \ . \tag{5-5}$$

This is the differential, projected solid angle. It can be rewritten as

$$d\Omega_{p1} = n^2 d\Omega_{p2}. \tag{5-6}$$

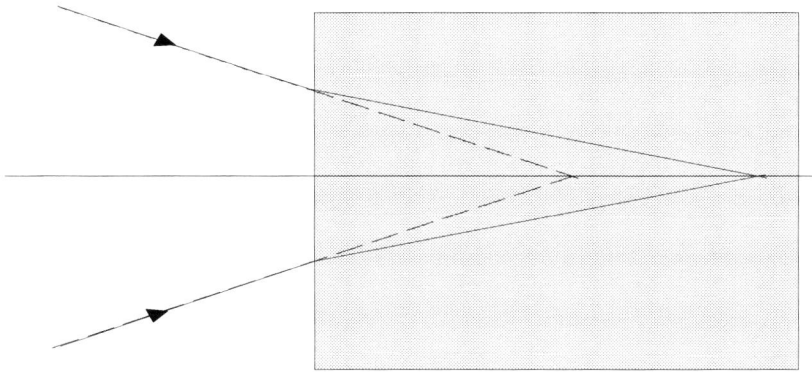

Figure 5-2. Conservation of radiance across an interface.

Since the refractive index of a transparent material must always be greater than 1, the solid angle in the material is always smaller than the solid angle in the vacuum and smaller by the square of the refractive index. This leads to a new form of invariance. The reduced radiance, defined as L/n^2, is a constant throughout a system, no matter what the material. The fundamental equation of radiative transfer can then be written

$$d\Phi_1 = L_1 A_1 d\Omega_{p1} = L_2 A_1 d\Omega_{p2} = L_2 A_1 \frac{d\Omega_{p1}}{n^2}. \qquad (5\text{-}7)$$

So

$$L_1 = \frac{L_2}{n^2}. \qquad (5\text{-}8)$$

The invariant is the radiance divided by the square of the refractive index. Some investigators call this the reduced radiance or basic radiance. The throughput also changes by the square of the refractive index, but in the opposite direction so as to maintain the constancy of power.

This same result can be obtained in another way. The expression for blackbody spectral radiance is

$$L_\lambda(\lambda, T) d\lambda = \frac{2hc^2}{\lambda^5 (e^{hc/\lambda kT} - 1)} d\lambda, \qquad (5\text{-}9)$$

where λ is wavelength of light in the material, T is absolute temperature, h is Planck's constant, c is the speed of light, and k is the Boltzmann constant. The relationship between the wavelength in the medium, λ, and that in a vacuum, λ_0, is n, that is, $\lambda_0 = n\lambda$, and therefore the ratio of radiance in a medium to that in a vacuum is

$$\frac{L_\lambda(\lambda, T) d\lambda}{L_{\lambda_0}(\lambda, T) d\lambda_0} = \frac{\dfrac{2hv^2}{\lambda^5 (e^{hv/\lambda kT} - 1)} d\lambda}{\dfrac{2hc^2}{\lambda_0^5 (e^{hc/\lambda_0 kT} - 1)} d\lambda_0} = n^2. \qquad (5\text{-}10)$$

Therefore, the fundamental equation might be written

$$\Phi = \frac{L}{n^2} n^2 A \cos\theta\Omega = L_r A\Omega_p = L_r Z, \qquad (5\text{-}11)$$

where L_r is the reduced radiance and Ω_p is the projected solid angle.

The radiance will be reduced by transmission losses that are the result of either scattering or absorption. These losses are taken into account by the transmission factor, usually written as τ, so that the transfer equation in a medium that refracts, reflects, absorbs, and scatters can be written

$$\Phi = \tau L_r Z. \qquad (5\text{-}12)$$

5.3 THE INVARIANCE OF THROUGHPUT

In a sense this proof can be obtained by backing up. We know that the power, which is invariant and must be conserved, can be written

$$\Phi = L_1 A_1 \cos\theta\Omega_{p1} = L_2 A_1 \cos\theta\Omega_{p2} = \frac{L_1}{n_1^2} Z_1 = \frac{L_2}{n_2^2} Z_2 = L_{r1} Z_1 = L_{r2} Z_2. \qquad (5\text{-}13)$$

Since the reduced radiances are equal, the throughputs are equal. There is an alternate way of proving this using the concepts of ray tracing in optical systems. The result is the so-called Lagrange or Helmholtz relation,

$$n_1 y_1 \cos\theta_1 = n_2 y_2 \cos\theta_2, \qquad (5\text{-}14)$$

where n is the refractive index, y is the height of a ray and θ is the angle the ray makes with the optical axis. Both sides of the equation can be squared and muliplied by π without disrupting the equality, and one has

$$n_1^2 y_1^2 \cos^2\theta_1 = n_1^2 A_1^2 \cos\theta\Omega_1 = Z_1 = Z_2. \qquad (5\text{-}15)$$

5.4 PATH RADIANCE

A medium that has appreciable absorptance or scattering along the path is sometimes described as having a path radiance, that is, the radiance per unit length of path. In an absorber, the radiance may either increase or decrease as the path length increases, depending on whether the material absorption or emission is greater. That depends on the temperature and the spectral region.

Path radiance also is important in highly scattering paths. In this case, for instance, the sun shines on the scatterers and there is more and more radiance with increasing path because it "picks up" the in-scattered flux. In the same atmosphere the radiance from a headlight will be reduced by the out-scatter.

5.5 RECAP

Very simple geometry has been used to show that two very useful invariants exist that can be used in the evaluation of the radiometry of optical systems. They are the radiance divided by the square of the refractive index and the throughput. It might be noted that the throughput is $n^2 A \Omega$. Path radiance is a useful concept for optical paths that have varying flux along the path, as with scattering and absortive media.

CHAPTER 6

SOURCES

Many measurements in radiometry start with the use of a source. It illuminates an optical system or a detector. It may be part of a measurement of reflectance or transmittance. Sources are part of radiometry.

A comprehesive list of sources and their properties is not possible in the scope of this text. This chapter is intended to provide an overview and perspective rather than handbook data. Two good references for more detail, and for more references may be consulted.[1]

Sources may be categorized in a number of ways, but surely they separate into laser and non-laser sources. Laser sources need to be described in terms of their wavelength of operation, power, and whether or not they are continuous. If they are not continuous, then their repetition rate and pulse characteristics are important. For both types, beam spread and lifetime are also important. The other sources consist of cavity radiators, arc lamps, incandescent (mostly tungsten) lamps, and special types.

6.1 LASER SOURCES

There are many types of lasers: gas, diode, dye, excimer, and more. They also operate either as fixed-wavelength sources or can be tunable. The latter differentiation is made here.

6.1.1 FIXED-WAVELENGTH LASERS

The type that comes immediately to mind is the ubiquitous helium-neon (He-Ne) laser, which is a cheap, stable workhorse for almost every laboratory. It can be operated at several wavelengths, but the most common is the red line at 0.6328 μm. In this mode, it has an output power from 1 to 10 milliwatts. This laser can also be operated at 1.15 and 3.5 μm but with reduced performance in terms of both stability and power output.

---

[1]LaRocca, A. J., *Chapter 2,* "Artificial Sources," in W. Wolfe and G. Zissis, Eds., *The Infrared Handbook*, Environmental Research Institute of Michigan, 1978 (available from SPIE); ibid. in J. Accetta and D. Shumaker, *The Infrared and Electro-Optical Systems Handbook*, Environmental Research Institute of Michigan and SPIE Press, 1993.

The next most common gas laser is probably the carbon dioxide (CO_2) laser. Although it can be tuned by varying the cavity length to operate at a number of lines from about 9.6 µm to 11.6 µm, it is most often used midway between those two, at 10.6 µm. The output power is a function of the length of the gain medium, i.e., the length of the tube; the most common versions have an output on the order of 10 watts. The laser is quite stable.

The neodymium yttrium aluminum garnet (Nd:YAG) solid state laser is also frequently seen in the lab. In its continuous mode of operation it emits at 1.06 µm with an output from 1 to 1000 watts. It can be frequency doubled so that it emits at 0.53 µm, but then yields only about 1 watt. In a pulsed mode, it emits at the same two wavelengths with 0.1 to 20 joules per pulse that lasts from 0.02 to 100 µs. Of course the pulse width and the energy are related. Neodymium can also be used in a glass matrix, and has similar characteristics.

The main diode lasers are GaAs and PbSnTe, although GaAsP and other mixed crystals have made possible diode lasers that are red. The gallium-type lasers typically radiate from 0.6 to 0.9 µm (at fixed wavelengths) with powers on the order of milliwatts. The PbSnTe is tunable and discussed in the next section. The diodes can also be pulsed with peak powers ranging from 1 to 2000 watts, pulse widths of about a millisecond, and repetition rates of one thousand to one million per second. The laser pointer is typical of some of these devices.

6.1.2 TUNABLE LASERS

There are several different types of tunable lasers, all with different characteristics[2]. The most useful seem to be semiconductor lasers[3] that can span the spectrum from about 350 nm to about 14 µm. This range must be accomplished with different materials ranging from ZnCdS to HgCdTe of various mixture ratios. Tuning is accomplished in these semiconductor lasers by varying either the drive current or the pressure. Powers range from microwatts to milliwatts; line widths as small as 10^{-6} cm^{-1} have been observed, but useful widths are somewhat larger. Note that this represents a resolving power of about 10^{10} !

---

[2]Weber, M. J., ed., *CRC Handbook of Laser Science and Technology*, CRC Press, 1986.

[3]Koechner, W., *Solid State Laser Engineering*, Springer-Verlag, 1976.

The second most useful tunable laser is probably the family of dye lasers[4]. These cover the region from about 350 nm to 1 μm using several different dyes and solvents. Each combination can be tuned several hundred wavenumbers from its spectral peak with a concomitant reduction in power.

Color-center lasers, in which lasers like Nd:YAG pump certain alkali halides, cover the region from about 1 μm to 3 μm, again with several different pumps.

The fourth class is molecular lasers like carbon dioxide, carbon monoxide, deuterium fluoride, and carbon disulfide. These molecular lasers are used at relatively high pressures to broaden the vibrational-rotational bands of the gases themselves. These typically have much greater flux, but somewhat lower resolving power.

A summary of the properties of the main tunable lasers is given in Table 6-1.

6.2 BLACKBODIES

A real blackbody radiator is a figment of a theoretician's imagination. It is the perfect, equilibrium radiator that was considered by Wien[5], Rayleigh[6], Jeans, Stefan, Boltzmann, and finally Planck. Their theoretical considerations led to the quantum theory, first proposed, somewhat uncomfortably, by Planck and refined by Einstein and others. Although the history and the derivations are both fascinating, only the significant results will be presented here. The expression for the spectral radiant exitance, M_λ, is

$$M_\lambda = \frac{c_1}{\lambda^5 (e^x - 1)}, \qquad (6\text{-}1)$$

where c_1 is the first radiation constant, λ is the wavelength, x is $c_2/\lambda T$, the dimensionless frequency, c_2 is the second radiation constant, and T is the absolute

---

[4]Schaefer, F. P., "Principles of Dye Laser Operation," in F. P. Schaefer, ed., *Dye Lasers, Springer-Verlag, 1973.*

[5]Wien, W., Sitzungsberichte der Akademie der Wissenschaften, Berlin 1893.

[6]Rayleigh (John William Strutt), Philosophical Magazine **49**, 539, 1900.

Table 6-1. Properties of tunable lasers.

Semi-conductor Lasers	Spectral Range [μm]	Dye Lasers	Spectral Range [μm]	Molecular Lasers	Spectral Range [μm]	Color-Center Lasers	Spectral Range [μm]
ZnCdS	0.3-0.5	Stilbene	0.4-0.42	DF	3.8-4.2	NaF	1.0-1.2
CdSeS	0.5-0.7	Coumarin	0.42-0.55	CO	5.0-8.0	KF	1.2-1.6
GaAlAs	0.6-0.9	Rhodamine	0.55-0.7	CO_2	9.0-12.0	NaCl	1.4-1.8
GaAsP	0.6-0.9	Oxazine	0.7-0.8	CS_2	9.0-12.0	KCl	1.8-1.9
GaAs	0.8-0.9	DOTC	0.75-0.8			KCl:Na	2.2-2.4
GaAsSb	1.0-1.7	HIDC	0.8-0.9			KICl:Li	2.3-2.8
InAsP	1.0-3.2	IR-140	0.9-1.0			RbCl:Li	2.8-3.2
InGaAs	1.0-3.2						
InAsSb	3.0-5.5						
PbSSe	4.0-8.0						
HgCdTe	3.2-15						
PbSnTe	3.2-15						

temperature. The temperature in this text is given in kelvins. The values of c_1 and c_2 are, respectively, $2\pi c^2 h = 37418.44$ [W cm^{-2} μm^4 and $hc/k = 1.438769$ [cmK]. This curve is shown as Figure 6-1.

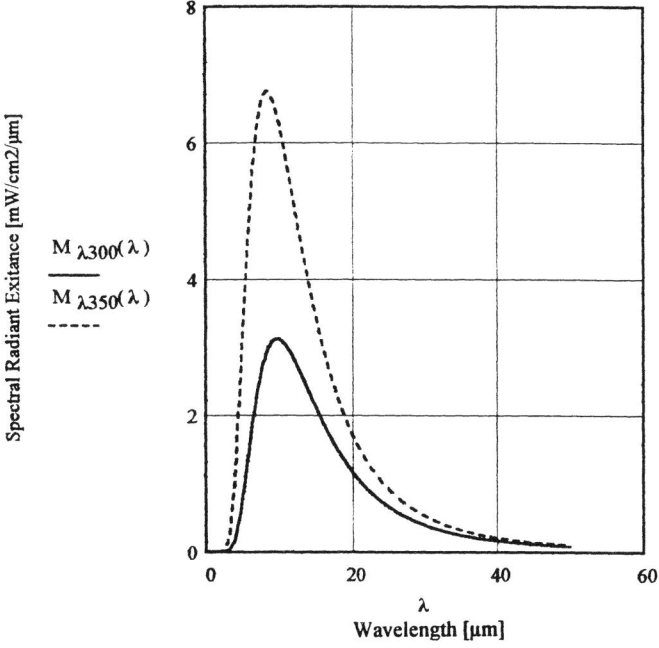

Figure 6-1. Spectral radiance exitance versus wavelength for 300 and 350 K.

There is an equivalent expression for the spectral radiant exitance, M_σ. The process of getting to that expression is to write

$$M_\sigma d\sigma = M_\lambda d\lambda \qquad (6\text{-}2)$$

rather than try to equate M_σ with M_λ, because Equation 6-2 equates flux densities in narrow spectral bands. Then,

$$M_\sigma = M_\lambda \frac{d\lambda}{d\sigma} = \frac{M_\lambda}{\lambda^2} = \frac{c_1 \sigma^5}{e^{x}-1} \ , \qquad (6\text{-}3)$$

$$\sigma = \frac{1}{\lambda} \quad so\ that \quad d\sigma = -\frac{d\lambda}{\lambda^2}. \qquad (6\text{-}4)$$

The minus sign is not used, since this is a metrical translation[7]. Figure 6-2 shows these curves, i.e., M_σ versus σ for a set of temperatures.

The integral of both of these equations can be shown to be given by the Stefan-Boltzmann law,

$$M = \int_0^\infty M_\lambda d\lambda = \int_0^\infty M_\sigma d\sigma = \sigma T^4 . \qquad (6\text{-}5)$$

where σ is the Stefan-Boltzmann constant $= \pi^5 k^4 / 45 c^2 h^3 = 5.66962 \times 10^{-8} [\text{W m}^{-2}\text{K}^{-1}]$. There should never be any confusion between the wave number, σ, and the Stefan-Boltzmann constant, σ. The Stefan-Boltzmann law can be used to show that a typical two-square-meter blackboard gives off a kilowatt of radiation. Most people do not appreciate this fact until they actually go through the calculation. It does have ramifications for radiometry: there is a tremendous amount of background radiation that can interfere with many experiments.

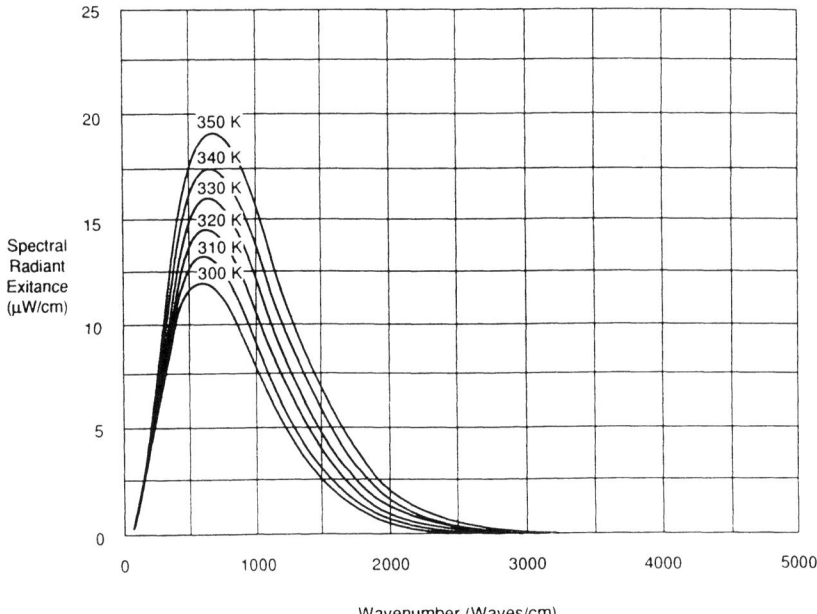

Figure 6-2. Spectral radiant exitance versus wavenumber.

---

[7]Frieden, B. R., *Probability, Statistical Optics, and Data Testing*, Second Edition, Springer Verlag, 1991.

The maximum of the M_λ vs λ curve is given by

$$\lambda_{max} T = 2897.756 \approx 3000 \, [\mu m K] \, . \tag{6-6}$$

The maximum of the M_σ vs σ curve is given by

$$\frac{T}{\sigma_{max}} = 5099.4149 \, [\mu m K] \, . \tag{6-7}$$

These equations can be used to find that the peak of the Planckian curve for ambient radiation is at about 10 μm. This emphasizes the fact that the background radiation, mentioned above, occurs largely in the infrared region of the spectrum, and is usually not a prbolem for visible measurements. The photon distributions can be obtained by dividing the expressions of Eqs. 6-1 and 6-3 by the energy of a photon to get

$$M_{q\lambda} = \frac{2\pi c}{\lambda^4 (e^x - 1)} \tag{6-8}$$

and

$$M_{q\sigma} = \frac{2\pi c \sigma^2}{e^x - 1} . \tag{6-9}$$

The maxima for these distributions are

$$\lambda_{max} T = 3669.72613 \, [\mu m K] \tag{6-10}$$

and

$$\frac{T}{\sigma_{max}} = 11076.30384 \, [\mu m K] \, . \tag{6-11}$$

The total photon radiation is given by

$$M_q = \sigma_q T^3 = \frac{2.4041(2\pi k^3 T^3)}{c^2 h^3} = 1.5202 \times 10^{11} T^3 \, [s^{-1} \, cm^{-2}] . \qquad (6\text{-}12)$$

This equation can be used to calculate the number of photons that are emitted from the aforementioned blackboard. The number is 4.104×10^{18}, or, as the late Carl Sagan would say, "billions and billions."

Of course, the power or the photons are not available for useful work, since they are in substantial equilibrium with the incoming radiation, and the second law of thermodynamics still applies. But, have no doubts, the photons and the watts can affect those infrared radiometric measurements.

The remaining equations can be used for the calculation of many different outputs. Just as lambertian surfaces are often used as first approximations for geometrical transfer calculations, blackbody expressions are often used for spectral flux output calculations.

6.3 CAVITY RADIATORS

These devices, sometimes called blackbodies, make use of multiple reflections in a cavity, with a little (or a lot) of absorption at each reflection to obtain a high emissivity. Consider a cavity that generates 110 reflections and is made of a material that has the relatively high reflectivity of 0.9. The result of all these reflections means that the effective reflectivity is less than 0.0001, meaning that the emissivity is 0.9999. Such cavities have been studied by many investigators, and there are several prescriptions for the design of them—at several different levels of complexity and accuracy. I prefer not to call them blackbodies, which, after all, is really just a theoretical concept, but instead *cavity radiators*, which is what most people mean when they say *blackbody*.

The simplest theory is for a cavity that is both uniform in temperature and lambertian in its reflectivity. It can be shaped as a circular cylinder, cone, cylindro-cone, or other, similar configuration. One simple rule seems to be that if you make the interior area large with respect to the aperture area, then the emissivity will be high. However, for commercial practice, one wants to minimize the size and weight, and therefore the shape and depth of the cavity are of the essence.

Gouffé[8] has presented a simple theory for calculating the emissivity of the aperture of a cavity of various shapes and materials. He states that the emissivity of the cavity is

$$\varepsilon_c = \left| \frac{\varepsilon}{\varepsilon\left(1 - \frac{s}{S}\right) + \frac{s}{S}} \right| \left[1 + (1 - \varepsilon)\left(\frac{s}{S} - \frac{s}{S_0}\right) \right],$$

(6-13)

where s is the area of the aperture, S is the area of the interior surface, S_0 is the area of the sphere that has a diameter equal to the depth of the cavity, and ε is the emissivity of the interior surface. Figure 6-3 shows a comparison between cylindrical and spherical cavities. Figure 6-4 shows a comparison of cylindrical cavities with different emissivities.

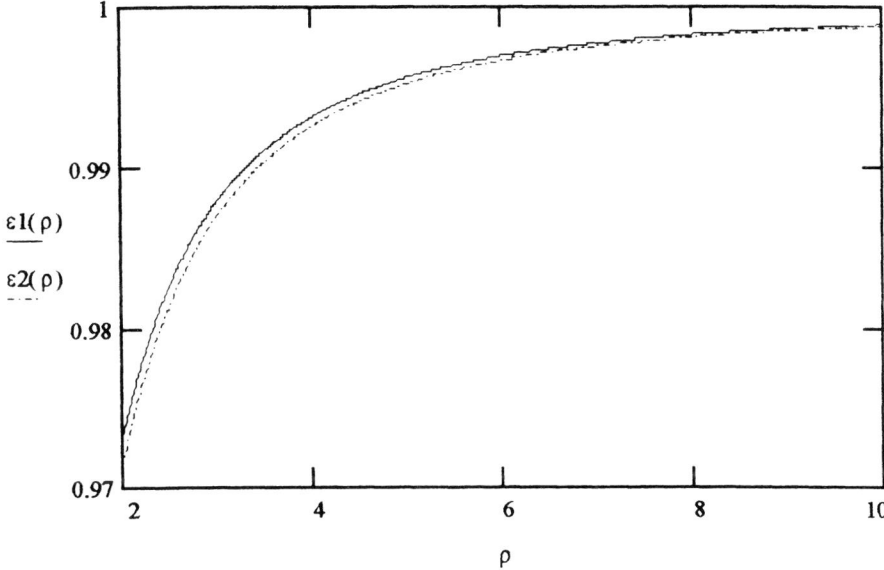

Figure 6-3. Cavity emissivity for spherical and cylindrical cavities as a function of depth-to-aperture ratio. The cylindrical cavity is a little worse than the spherical one.

[8]Gouffé, A., "Corrections d'ouverture des corps-noir artificiels compte tenu des diffusions multiples internes," Revue d'Optique **24**, 1, 1945.

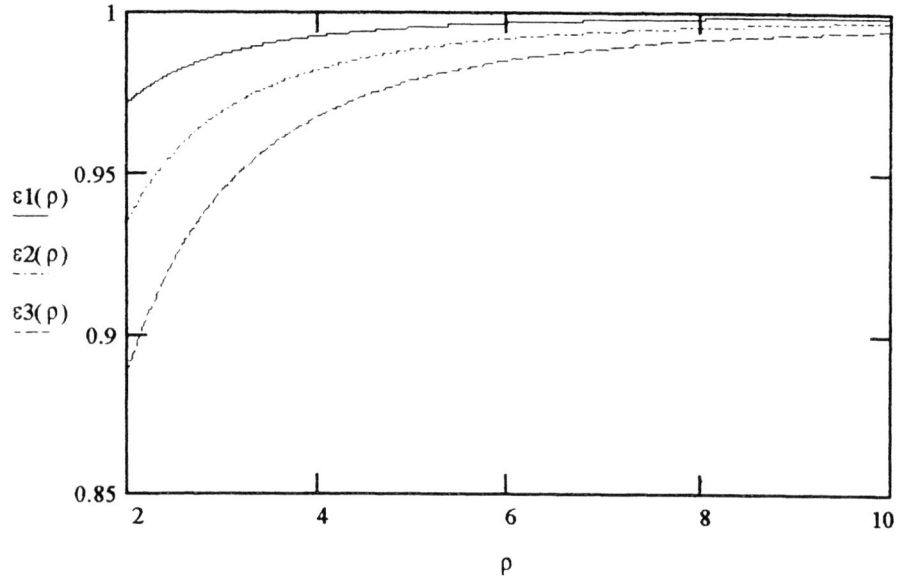

Figure 6-4. Cavity emissivity for cylindrical cavities as a function of depth-to-aperture ratio for emissivities of 0.9, 0.8, and 0.7.

DeVos[9] carried out a more complicated evaluation to obtain the following expression:

$$\varepsilon_c = 1 - \rho_{bd}(0,0)\frac{\pi}{a^2} - 2\pi^2 \int_0^a \frac{\rho_{bd}(0,y)\rho_y(w,0)}{(y^2+1)[(y-a)^2+1]^{3/2}} dy .$$ (6-14)

Figure 6-5 helps to define the symbols. Figure 6-6 shows a comparison between the two theories for the emissivity of an open-ended cylindrical cavity. There have been an almost uncountable number of articles on the calculation of the

[9]DeVos, J. C., "Evaluation of the quality of a blackbody," Physica **20**, 669, 1954.

emissivity of cavity radiators of various shapes. These are discussed in Bartel's dissertation[10] and a related publication[11].

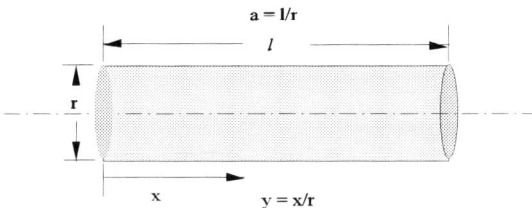

Figure 6-5. Symbol definitions for the DeVos calculation.

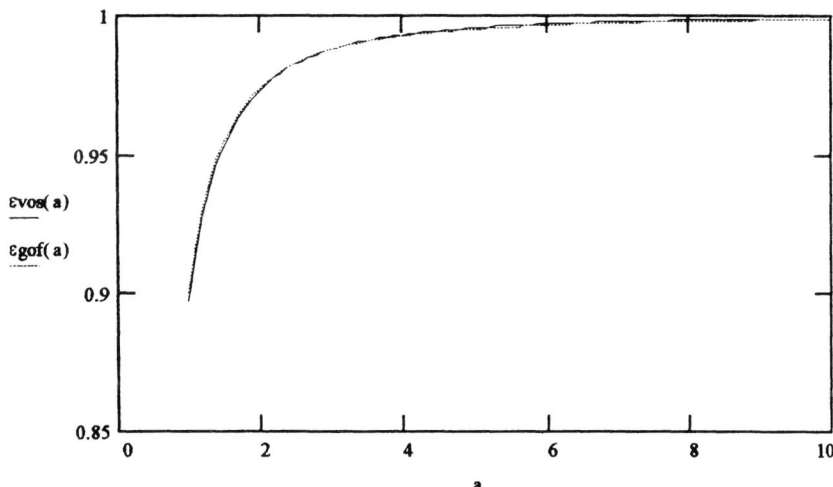

Figure 6-6. A comparison of the analysis of an open-ended cylinder by the methods of Gouffé and of DeVos.

[10]Bartel, F. E., "Blackbody Simulator Cavity Radiation Theory," Ph.D. dissertation, The University of Arizona, Tucson, AZ, 1975.

[11]Bartel, F. E. and W. L. Wolfe, "Cavity Radiators: an ecumenical theory," Applied Optics **15**, 84 1976.

Most commercial cavity radiators are cones or cylindro-cones that can be heated to about 1200°C. They are typically constructed of a ceramic core with heating coils around them and insulating materials around the heating coils. They heat up in a matter of minutes but take hours to cool down. The temperature calibration is usually quite good, but for the most accurate work, the radiator should be calibrated. One way is with a spectrometer to observe the spectral shape of the emission.

6.4 THERMAL SOURCES

The Nernst glower, Globar, Welsbach mantle, and tungsten-filament bulbs comprise this class of source. Although arcs of various sorts can be included, I classify them differently.

6.4.1 NERNST GLOWER

This source has long been used for spectrometer sources, largely because they are cylindrical in shape. They are made of refractory materials such as thoria, yttria, and zirconia and have an output that can be characterized fairly well as that of a 1600 K thermal radiator with a constant emissivity of about 0.75. Since the resistivity is higher at room temperature than at operating temperature, the glower must be heated to get it started, and it must have a ballast resistor to keep it from "running away," since it has a negative temperature coefficient of resistance. Since it is a thermal source, it is quite stable, the stability being largely determined by the stability of the power supply.

6.4.2 THE GLOBAR

This source is somewhat superior to the Nernst glower. It has a color temperature of about 1470 K, and comes in about the same sizes and shapes. It is a rod of bonded silicon with electrodes that must be cooled. It can be purchased commercially.

6.4.3 THE WELSBACH AND GAS MANTLES[12]

This is a radiator somewhat like that used in Coleman camping lanterns, but because of its relative instability is not a very good radiometric source. Other

---

[12]Ramsey, W. Y. and J. C. Alishouse, "A comparison of infrared sources," Infrared Physics **8**, 143, 1968.

variations are available[13].

6.4.4 TUNGSTEN BULBS

A wide variety of tungsten bulbs exist, having different sizes, shapes, forms, and envelopes. Tungsten is essentially a graybody operated at 2800 K with a total emissivity of about 0.34, but ranging up to 0.458 at 0.467 μm at 2800 K.[14]

6.5 RECAP

The choice of a source for a radiometric experiment can be difficult. On the one hand, there are so many different kinds with different properties. On the other, there never seems to be one that is just right. Considerations include the spectrum of operation, the required flux level, narrowness of beam, directivity, stability, sometimes coherence, and, of course, required power, weight, and cost.

---

[13]Pfund, A. H., "The electric Welsbach lamp," Journal of the Optical Society of America **26**, 439, 1936.

[14]*Handbook of Chemistry and Physics*, CRC Press, 1993.

CHAPTER 7

DETECTORS

Every radiometric measurement uses a detector in some way. This chapter discusses how they are described and delineates their most important properties with respect to radiometric measurements.

7.1 DETECTOR DESCRIPTIONS

The *responsivity* of a detector is the electrical output divided by the optical input. The inputs and outputs can take different forms. The output is often a voltage or a current, but sometimes it is a count of pulses or even a temporal frequency (which may have been generated by a voltage-to-frequency converter). In this text, for most purposes, the electrical output will be assumed to be either voltage or current. The input is a power or photon rate on the detector. It may be monochromatic, bandlimited, or cover the entire optical spectrum (say 95% to 99% of the radiation). The responsivity is almost always written as a script R, i.e., $\Re$.

Thus the signal voltage can be written as

$$V_s = \Re\Phi = \int \Re(\lambda)\Phi(\lambda)d\lambda. \qquad (7\text{-}1)$$

The total noise arises from several sources, but, no matter how many and what kind, the total noise can be written

$$V_n = \sqrt{\int V_n^2(f)df}, \qquad (7\text{-}2)$$

where the spot noise, the noise at a given frequency, $V_n(f)$, is the sum of all the noise components at that frequency. Therefore the signal-to-noise ratio, SNR, is given by

$$SNR = \frac{V_s}{V_n} = \frac{\Re\Phi}{V_n} = \frac{\int \Re(\lambda)\Phi(\lambda)d\lambda}{\sqrt{\int V_n^2(f)df}}. \qquad (7\text{-}3)$$

It is always nice to know the signal-to-noise ratio, but it does depend on the input signal and the bandwidth, the detector size, and the background. So, to compare detectors on an apples-to-apples basis, a specific detectivity has been defined. It is

$$D^* = \frac{\sqrt{A_d B}}{\Phi_d} SNR = \sqrt{A_d B} \frac{\Re}{V_n} . \qquad (7\text{-}4)$$

The specific detectivity may be thought of as a normalized signal-to-noise ratio, the SNR per unit power for a 1 cm^2 detector and 1 Hz noise bandwidth. The specific detectivity has the somewhat unusual units of cm $Hz^{1/2}$ W^{-1}.

7.2 DETECTOR TYPES

Radiation detectors may be categorized as those that detect photons and those that detect power. Somewhat more accurately, they are those that have a spectral responsivity that is independent of wavelength for a photon incidance and those that are independent for a power incidance. This arises from the basic mechanisms of detection.

Photon detectors operate when an electron is released from a bound state (the valence band) to a state in which it is free to participate in electrical conduction (the conduction band). Then, a photoconductor senses the resultant change in conductivity by using some sort of voltage divider network. A photodiode detector exhibits a change in the back-bias portion of its I-V (current-voltage) curve, as shown schematically in Figure 7-1. This change can be monitored a number of ways. A photovoltaic detector is a photodiode detector operated with no back bias.

Thermal detectors provide an output that is related to their temperature or change in temperature. The power to be detected is incident upon the surface, where it is absorbed. The absorbed power increases the temperature of the detector according to its mass and specific heat, and it loses heat according to the conduction path to a heat sink. When this heat is sensed as a change in resistance, the detector is called a bolometer. There are several kinds: metal, semiconductor, and superconductor. Two of the most frequently used are the thermistor bolometer, a *therm*ally sensitive res*istor* bolometer that is made of semiconductors of rare earth oxides, used mostly at room temperature, and the germanium bolometer, used principally at low temperatures (about 4 K). *Thermocouples* are pairs of junctions of dissimilar metals. Because of this difference, there is a potential difference generated in each of the junctions. If one is kept at a reference temperature, the other is a measure of the temperature it experiences. A *thermopile* is a series connection of a number of thermocouples. A *pyroelectric detector* is composed of a material with an asymmetrical crystal structure. It has a certain distribution of internal electric field and charges. A change in temperature causes a change in this distribution that can be sensed as a change in electric field, potential, or voltage.

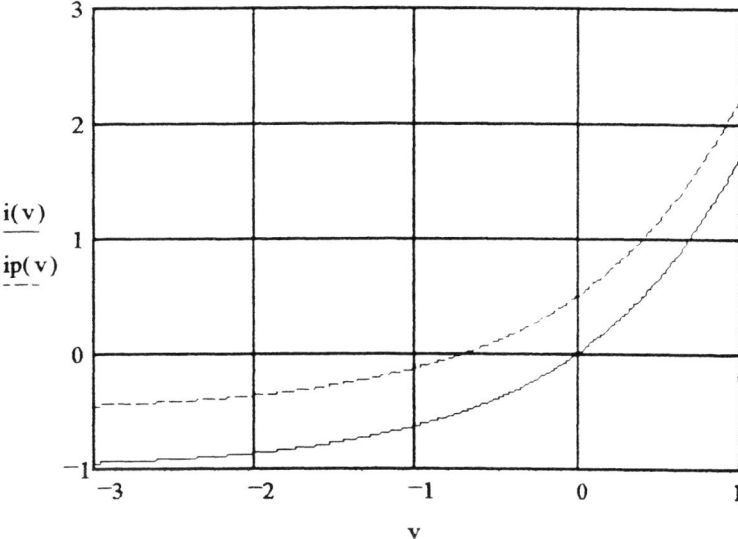

Figure 7-1. The I-V curve of a p-n junction. Increases exponentially with positive bias and flattens out with negative bias. The upper curve was caused by an increased incidance of photons.

Figure 7-2 shows idealized curves for the two types of detectors. The thermal detector is "flat," i.e., has a power responsivity that does not change with wavelength. The photon detector has a linear increase of its power responsivity with wavelength. If one wishes to measure the power from some source, a thermal detector is the one to use. As shown in Chapter 9 on normalization, one cannot actually measure the power in a band with a photon detector (unless the source spectrum is already known).

7.3 DETECTOR NOISES

There are six different types of noise that afflict the performance of radiometric detectors. These are Johnson noise, sometimes called thermal noise; shot noise, sometimes called Schottky noise; photon noise; temperature noise; generation-recombination (gr) noise; and excess noise, often called 1/f ("one over f") noise.

Every noise is a statistical variation around some average value, and is usually specified as the root-mean-square value. This indicates that one finds an average, averages the squares of the differences of each of the values from the average, and takes the square root of that average.

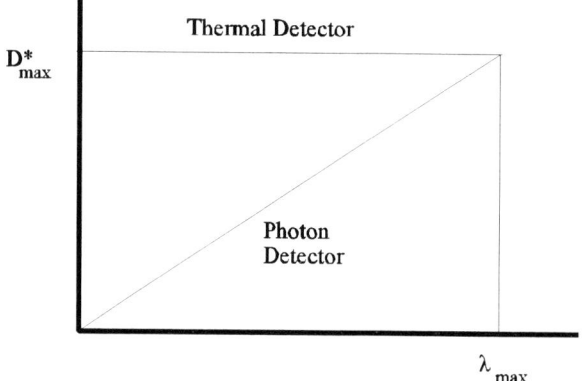

Figure 7-2. Idealized specific detectivity curves for photon and thermal detectors.

7.3.1 JOHNSON NOISE

This may be thought of as the fluctuation in the length of time it takes for carriers to effectively traverse a resistor. It only occurs in a resistive element and only when there is a current flowing. The mean-square (square of the rms) current is given by

$$i_{ms} = \frac{4kT}{R} B,$$ (7-5)

where k is Boltzmann's constant, T is the absolute temperature, R is the resistance, and B is the temporal bandwidth.

7.3.2 SHOT NOISE

This noise is associated with the fluctuation in the time it takes to surmount a potential barrier. The most vivid example of this is the fluctuation in emission times in a vacuum tube. For those of you who don't remember vacuum tubes, this may not help much. Think of it as the fluctuation in the time it takes to cross the potential barrier of a *p-n* junction. The expression for shot-noise mean-square current is

$$i_{ms} = 2qIB,$$

(7-6)

where q is the charge on the electron, I is the dc current, and B is again the temporal bandwidth.

7.3.3 PHOTON NOISE

Photons are "clumps of light." They arrive like rain on a detector, and there is both an average rate and a fluctuation around that average. The expression for the mean-square current is

$$i_{ms} = 2q\eta \bar{N}B,$$

(7-7)

where η is the quantum efficiency, the number of electrons produced per photon (less than 1), and N is the average photon rate. It is interesting that the average current and the mean-square deviation from the mean current are the same. Therefore the SNR for a photon stream is the square root of the average rate. This noise occurs only in photon detectors.

7.3.4 TEMPERATURE NOISE

Temperature noise may be considered the equivalent of photon noise, but for thermal detectors. It is the fluctuation in the detector temperature around the average temperature that arises because there is a fluctuation in the arrival of the photons that carry the energy. This noise occurs only in thermal detectors.

7.3.5 GENERATION-RECOMBINATION NOISE

This noise is related to the fluctuation around the average rate of generating carriers from the bound state and the fluctuation in their recombination rates. It occurs only in photon detectors, and its expression is

$$i_{ms} = \frac{2I^2\tau}{1+\omega^2\tau^2},\tag{7-8}$$

where I is the average current, ω is the temporal circular frequency, and τ is the detector time constant.

7.3.6 EXCESS NOISE

In a very real sense, this is what's left. It is not a fudge factor, however. It just isn't understood very well. Some experiments have shown it to be related to poor contacts; others, to poor surface states; still others to impurities in the volume or structure of a detector. But it is always there to some extent. The expression for excess noise current is

$$i_{ms} = \frac{aI^b}{f^c},\tag{7-9}$$

where I is the dc current; f is the temporal frequency; and a, b, and c are empirical constants, with c about equal to 2. Obviously, with all those empirical constants it really is not understood very well. It is an important noise, as it affects things at dc, and it causes the use of modulators and other manipulations.

7.4 SUMMARY OF NOISES

Figure 7-3 shows an idealized version of all these noises on the same chart. It is only representative, and does not indicate that one noise type is always larger than another. That depends on the detector and the photon flux on the detector. It does show that every noise is flat (independent of temporal frequency) except gr and excess noise.

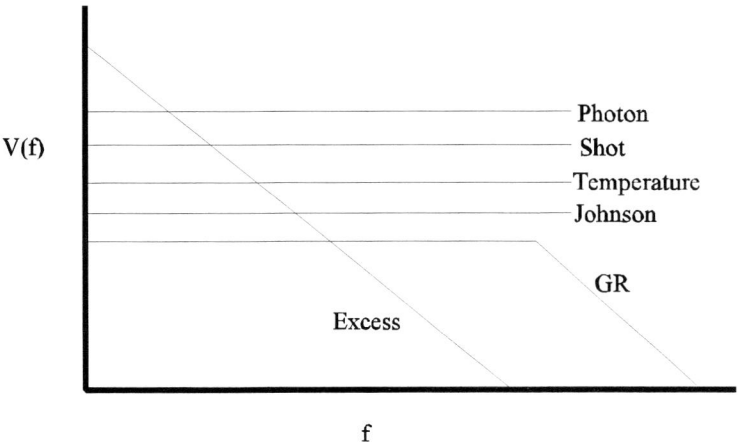

Figure 7-3. Schematic summary of noise spectra.

7.5 SUMMARY OF DETECTOR PROPERTIES

Table 7-1 provides a summary of the detectivities, spectral regions, time constants, and operating temperatures for the most important radiometric detectors.

Table 7-1. Radiometric detector properties.				
Detector	Spectrum [μm]	D^* [cm Hz$^{1/2}$ W^{-1}]	Time Constant [μs]	Temperature [K]
Si	0.5-1.1	5×10^{12}	0.1	300
InSb	1-5	10^{11}	0.1	77
HgCdTe	1-14	3×10^{10}	0.1	90-250
Thermistor	black	6×10^{8}	1000	300
Ge	black		1000	4
Pyroelectric	black	10^{9}	1000	300
Thermocouple	black	10^{9}	1000	300

Of course, these are representative values. The "pyros," thermistors, and thermocouples can trade sensitivity for time constant by altering the mass and the conduction path to the heat sink. Similar operations can be applied to the photon detectors by altering bias voltages and mobilities, via impurities. The germanium bolometer, which has often been the choice for long-wave detection in semiconductor labs and by astronomers, can be operated down to about 1 K. The cold bolometer does not have a specific detectivity listed. Most reports on this detector give the noise equivalent power (in watts per root Hertz) and do not give either the specific detectivity or the detector area. Low[1] reported a good value of 5×10^{-13} W $Hz^{-1/2}$. If the detector had an area of 1 cm^2, it would have a specific detectivity of 5×10^{12}. The area was probably smaller, and the specific detectivity larger.

7.6 SOME REAL PROBLEMS

Most of the above discussion has been about idealized detectors. Like men (and women), almost none works like the ideal. They have variations in sensitivity over their areas. Not all of the same type have the same (spectral) curves. They can be inconsistent from day to day. These are considerations that must be made in their use as radiometric transducers.

7.7 RECAP

The nature of the two major types of detectors, photon and thermal, has been outlined. Their noise sources have been described and their performance characteristics outlined briefly. Certain detectors, notably photomultiplier tubes and doped silicon and germanium devices, have been omitted largely because they do not make good radiometric transducers. The reader is well advised to check with a supplier on such properties as dynamic range, stability, lifetime, and reproducibility, as well as the properties described above—detectivity, time constant, spectral coverage, and uniformity.

---

[1]Low, F. J., "Low temperature bolometer," *Journal of the Optical Society of America* **51**, 1300 (1961).

CHAPTER 8

REVIEW OF OPTICS [1]

Radiometry deals with calculating or measuring a quantity of radiation, whether in optical systems, in the atmosphere, on the Earth, on planets, or elsewhere. This chapter reviews some of the principles of optics that are essential and even useful in radiometry.

8.1 PHOTONS, WAVES, AND RAYS

Radiation can be described in terms of rays, waves, photons, and even probability amplitudes. The latter is not necessary in radiometry, but an interesting use of it is described by Feynman[2]: the derivation of the law of reflection.

Light is electromagnetic radiation of the same type as fm, am, and short-wave radio waves, television, and X-rays. They can be described mathematically by a wave function Ψ. A traveling wave will be a function with an argument of $t - z/v$, where t is time, z is the direction of travel, and v is the velocity of travel.

Light has also been shown to consist of photons, small clumps of energy, hc/λ. This model is particularly suitable in analyzing the detection of radiation.

One can visualize waves by analogy to water waves in a pool. When a rock is dropped in a placid pool, waves arise. The crests expand in circles away from the point of entry. Each crest represents a wavefront. Each trough also represents a wavefront. These wavefronts are circles, the perimeters of circles. The rays are normal to them, the radii of the circles. The direction of the travel of light can be represented by the rays, the normals to wavefronts.

The reader should realize that these are all just *models* of radiation. Each is appropriate for describing a certain application or situation. An apt model is one

---

[1]Born, M. And E. Wolf, *Principles of Optics*, Pergamon, 1959; Jenkins, F. and H. White, *Fundamentals of Optics*, McGraw Hill, 1957; S. G. Lipson, H. Lipson, and D. Tannhauser, *Optical Physics*, Cambridge University Press, 1995; B. Rossi, *Optics*, Addison Wesley, 1957.

[2]Feynman, R., *QED, The Strange Theory of Light and Matter*, Princeton University Press, 1985.

that is as simple as it can be to describe a phenomenon to the required accuracy. The photon model, for instance, is used here to describe photon noise. The wave model is used to describe interference and diffraction. The ray model is used to describe lens action and aberrations.

8.2 INTERFERENCE

The expression, from electromagnetic theory, for the incidance, the flux density, is given as

$$E = \frac{1}{2}\eta\Psi\cdot\Psi^* ,\qquad (8\text{-}1)$$

where η is the admittance of free space, a constant, and Ψ is the wave function. The asterisk indicates the complex conjugate. When two waves combine, one has

$$E = \frac{1}{2}\eta(\Psi_1+\Psi_2)\cdot(\Psi_1+\Psi_2)^* = \frac{1}{2}\eta(\Psi_1^2+\Psi_2^2+2\Psi_1\Psi_2\cos\varphi) ,\qquad (8\text{-}2)$$

where the subscripts indicate wave one and wave two, ω is the radian frequency, t is time, and φ represents the phase difference between the waves,

$$\varphi = \frac{2\pi}{\lambda}nd\cos\theta ,\qquad (8\text{-}3)$$

where λ is the wavelength, n is the refractive index, d is the distance between the waves, and θ is the angle between them.

It can be seen that two, equal-amplitude, monochromatic waves that interfere will generate a pattern characterized by a fixed term and a cosinusoidal one, as shown in Figure 8-1.

8.3 DIFFRACTION

Diffraction is generated by apertures and obstacles that limit the lateral extent of a wave. The Huygens-Kirchhoff theory represents this as the interference of an infinite number of daughter wavelets. The equation for such diffraction (in which all distances are large compared to the wavelength of light and caused by a rectangle) is

$$E = E_0 sinc^2(ka\cos\theta)sinc^2(kb\cos\theta) \, , \qquad (8\text{-}4)$$

where E is the flux density, E_0 is the initial flux density, a is one side of a rectangle, b is the other side, k is $2\pi/\lambda$, and the sinc function is the sine divided by its argument, i.e., $sinc(x) = (\sin x)/x$. The pattern of Equation 8-4 is shown in Figure 8-2. It is a little different for a circular aperture:

$$E = E_0 jinc^2(kr\cos\theta) \, , \qquad (8\text{-}5)$$

where r is the radius of the aperture and the jinc function is the Bessel function of the first kind divided by its argument, $J_1(x)/x$.

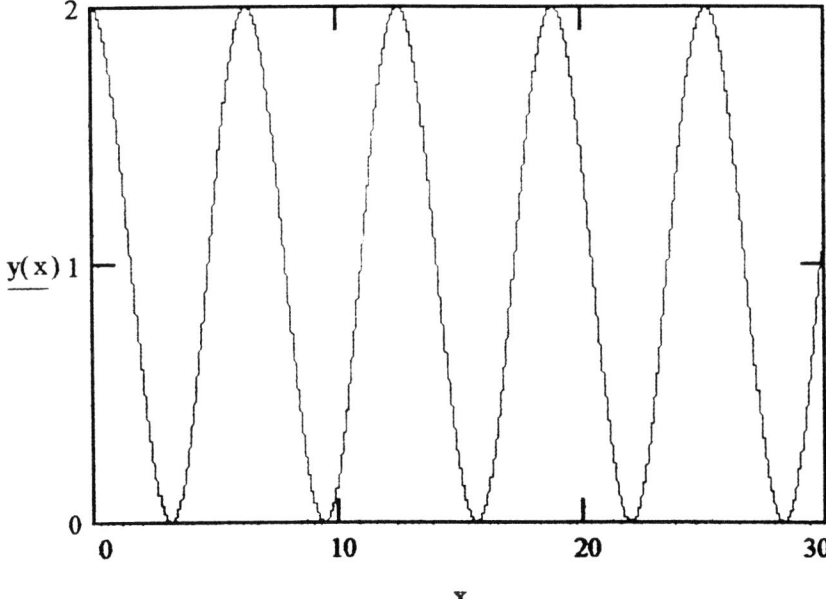

Figure 8-1. A typical interference pattern.

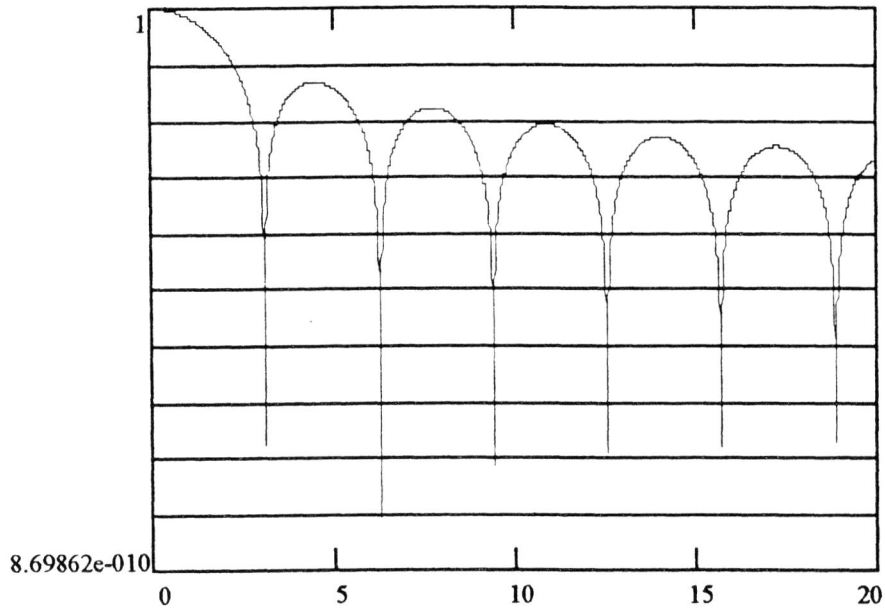

Figure 8-2. A diffraction pattern plotted logarithmically.

8.4 THE THIN LENS

The simplest model of a lens is the so-called thin lens using gaussian optics. It is illustrated in Figure 8-3, and provides the well-known equation for the image distance based on the object distance and the focal length,

$$\frac{1}{f} = \frac{1}{o} + \frac{1}{i} , \qquad\qquad (8\text{-}6)$$

where o is the object distance, i is the image distance, and f is the focal length. The back focal length is defined as the point on the optical axis where a ray from infinity (a ray parallel to the optical axis) crosses the axis. The front focal length is the point of axis crossing for a parallel ray from image space. Equation 8-6 is derived by relatively simple plane geometry.

8.5 RAY TRACES

For the relatively simple case of a spherical lens, several equations can be used to see what happens to a ray that is incident at a given height and angle. Similar techniques can be used for aspheric surfaces.[3]

These are called the refraction and the transfer equations, and they can be used in succession to determine the ray path through very complicated optical systems. This is ray tracing.

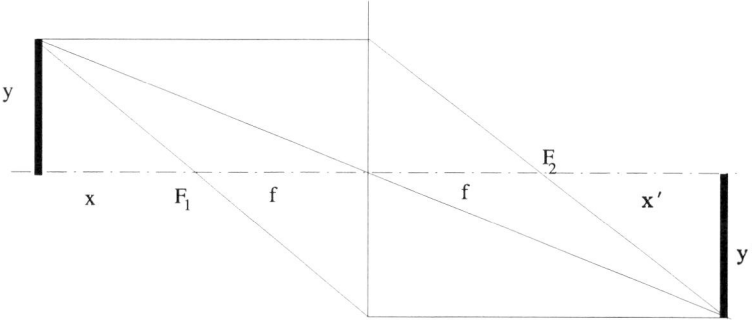

Figure 8-3. The thin lens. F_1 and F_2 are the front and rear focal points; f is the focal length, y and y' are the object and image heights, and x is the distance from the object to the focal point.

The differences between calculations based on true ray tracing and various approximations, like the use of angles for sines, are called aberrations. Recall that the sine function can be represented by an infinite series of odd terms:

$$\sin(x) = x + \frac{x^3}{3!} + \frac{x^5}{5!} \cdots + \cdots +$$

(8-7)

Thus, one can have third-order, fifth-order and higher-order aberrations. For radiometry it is important to know that these imperfections and corrections to gaussian optics exist. They can influence measurements if not handled correctly.

---

[3]Smith, W. J., "Optical Design," in W. Wolfe and G. Zissis, *The Infrared Handbook*, US Government Printing Office, 1985 (available from SPIE).

8.6 PARAXIAL RAY TRACES

An approximation to these rather tedious, ray-trace calculations, which is done today with one of many different, commercially available computer programs, is the so-called paraxial approximation. This approximation assumes that the rays are close enough to the optical axis that sines and tangents can be approximated by their angles. This is, in fact, a linearization of the ray-tracing equations, and is an approximation.

8.7 ABERRATIONS

If one improves paraxial calculations by adding the next term of the series representation for the sine and tangent, and calculates the difference, the third-order aberrations are obtained. They are so called because the first term is proportional to the angle, and the second term is proportional to the cube (third power) of the angle. The third-order aberrations, also called Seidel aberrations after one of the early investigators, are spherical aberration, coma, astigmatism, curvature of field, and distortion. Lateral and longitudinal chromatic aberration or color are usually considered with them.

These aberrations in general combine in any optical system to cause an object point to become an image blur. In most of radiometry, these effects are ignored. That is, the calculations for image incidence assume that gaussian (linear, one-to-one) optics is valid. In particular, the flux density falls off as the fourth power of the cosine of the field angle in an idealized approximation. The fall-off expression must be modified by the fact that the blur circle increases with field angle due to coma, astigmatism, curvature of field, and maybe distortion.

8.7.1 SPHERICAL ABERRATION

For most practical arrangements of the object and the lens, spherical surfaces do not bring the rays from an infinitely distant, on-axis object to a point on the axis in image space. The general arrangement is shown in Figure 8-4. This aberration is independent of field angle; it occurs on axis as well as at finite field angles.

8.7.2 COMA

This is an off-axis aberration that arises as an object point moves off axis. The rays reach the lens from different angles and therefore "see" a different thickness of lens and an apparently different curvature, so they have a varying magnification. The general appearance of the image of an object point is a comatic blur, as shown in Figure 8-5.

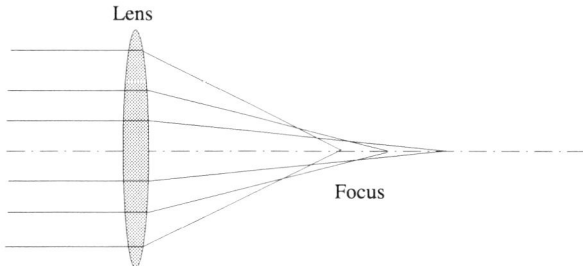

Figure 8-4. Illustration of exaggerated spherical aberration.

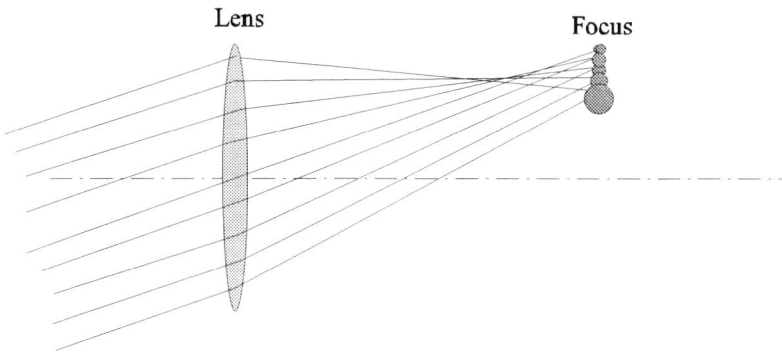

Figure 8-5. Illustrating comatic aberration.

8.7.3 ASTIGMATISM

At angles even further off axis the effect becomes even more pronounced—to the extent that an object point has two separated linear images that are, respectively, vertical and horizontal, with a best-focus image in between the two. Astigmatism is portrayed schematically in Figure 8-6. Most systems operate at best focus. The rays in the tangential plane come to a nice line focus at the tangential focus, indicated by the T in the figure. The sagittal rays do the same at S.

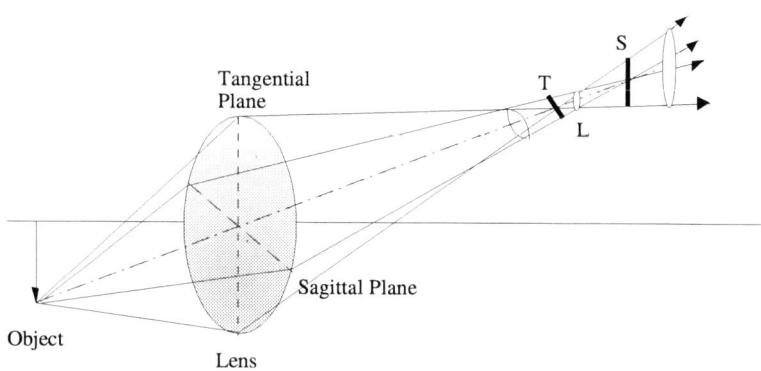

Figure 8-6. Illustrating astigmatism. The object is in the plane of the paper. The tangential plane of the lens gives sharp images at T; the sagittal at S; the circle of least confusion is at L.

8.7.4 CURVATURE OF FIELD

Because the various surfaces of a complicated lens are all curved, the image surface that provides the best imagery is a curved surface. In the course of design, a balancing of convex and concave surfaces can help to minimize this curvature. Because film and CCD arrays are flat, the blur one gets on a plane as opposed to the natural curvature is considered an aberration. Curvature of field is shown schematically in Figure 8-7.

8.7.5 DISTORTION

Things just ain't linear. A fixed grid in object space is not imaged as a fixed grid in imaging space. The metrics are a little different. A rectangular grid will be imaged as either a pincushionor a barrel.

8.7.6 LONGITUDINAL COLOR

Color, or chromatic, aberrations are caused by the fact that the refractive index is a function of the wavelength of the light. The power of the lens is proportional to the refractivity (n-1), so an object point is imaged at different points on the axis.

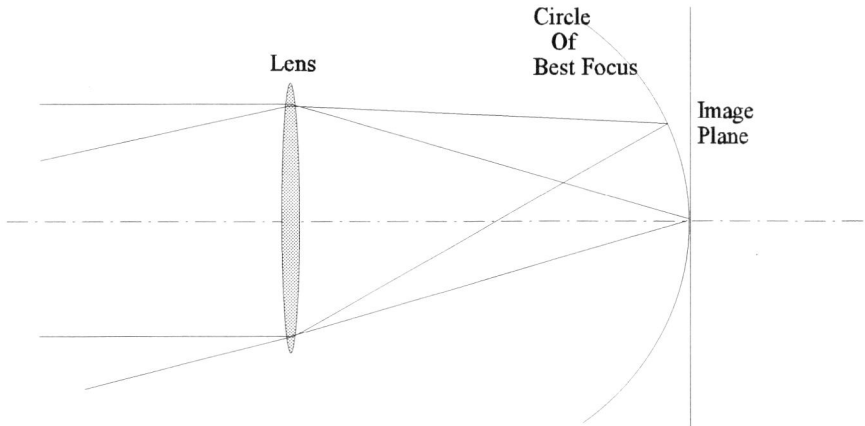

Figure 8-7. Illustrating curvature of field (curvature of the image surface).

8.7.7 LATERAL COLOR

The same effect can be seen at an image plane that is perpendicular to the optical axis. The object point will be a perfect, on-axis point for some wavelength, and the others will be blurs of various radii at the same image plane.

8.8 STOPS AND PUPILS

Stops provide physical limits to the size of the beam of light that can be accepted and the field of view that can be covered. Pupils are images of stops.

8.8.1 APERTURE STOPS AND PUPILS

The aperture stop limits the size of the beam of radiation that can be accepted by the optical system of a radiometer. Figure 8-8 shows a single lens performing as a stop. The beam size is determined by the diameter of the lens on axis, and off axis is reduced by the cosine of the field angle. Thus, the rim of the lens is the aperture stop. This is completely straightforward with such a simple system. The entrance pupil is defined as the image of the aperture stop that is formed by all elements that precede it. In this case, there are no preceding elements. The entrance pupil is identical to the aperture stop. The exit pupil is defined as the image of the aperture stop formed by all succeeding elements. It is also identical to the aperture stop. The lens is aperture stop, entrance pupil, and exit pupil.

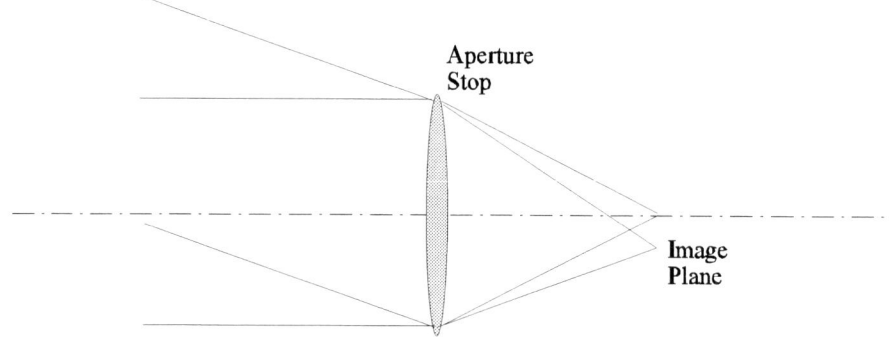

Figure 8-8. Illustrating the operation of an aperture stop.

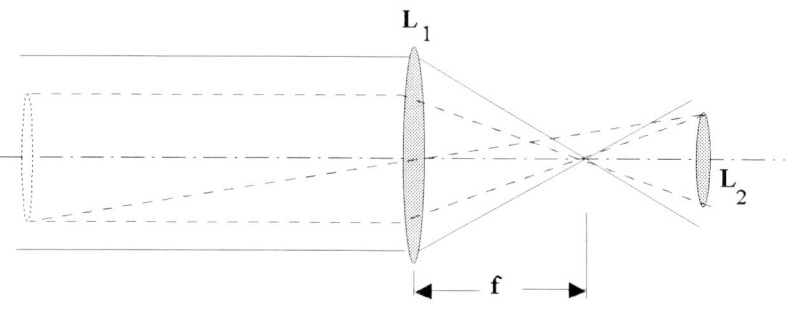

Figure 8-9. A two-lens system illustrating the determination of the aperture stop and the entrance pupil.

Figure 8-9 shows a two-lens system, in fact, a Keplerian telescope. Parallel rays come from infinity and form a point image at the focal point of lens L_1, designated as f. They diverge to lens L_2, which, as shown, is too small to collect them all. It is clear that the second lens is therefore the aperture stop. The entrance pupil is its image as shown by the dashed lines through the focus and the center of the lens. This is not a good design; it is better to have the image of the second lens either just match the first one or be a little oversized. A similar calculation can be made for the exit pupil.

8.8.2 FIELD STOPS AND WINDOWS

The field stop determines the angular extent from which rays may enter and be used by the system. It is usually located at or very near a focal plane. The entrance window is the linear field of view in object space. Figure 8-10 illustrates these concepts. The exit window is the linear size of the field stop in image space. The system shown is simple. When more complicated systems are used, there may be a number of candidate field stops. The determination of the real field stop is similar to that of determining the aperture stop.

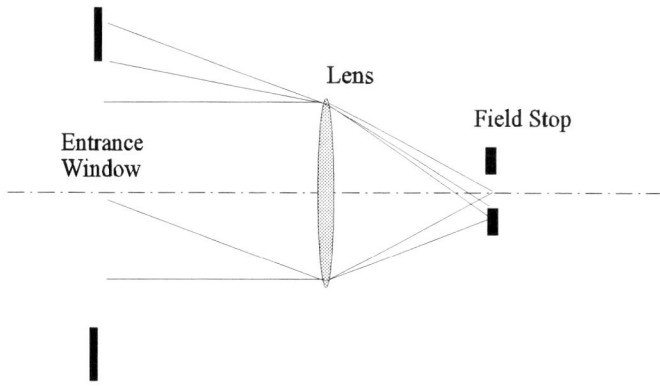

Figure 8-10. Field stop and entrance window.

8.9 RECAP

This chapter has touched very lightly on some general optics principles that bear on radiometry. Generally, interference and diffraction are not much considered in radiometry, but, as shown, they can have an influence on measurements. These phenomena can have particularly important effects with monochromatic and coherent radiation. Aberrations are normally ignored, with the caveat that radiometry deals with first-order optics. The investigator who ignores their influences, however, does so at his or her peril. Aberrations can do some obvious things like spreading the flux outside the detector sensitive area or outside the field or aperture stop.

Chapter 9

Normalization

The process of normalization is to bring calculations and measurements into consonance with a particular norm. One example is to relate everything to the response of the human eye. It has the advantage of making many calculations and measurements much easier in a particular application, but also has some very subtle and difficult pitfalls. This chapter discusses the process, gives examples, and shows the pitfalls.

9.1 The Need for Normalization

There is no such thing as a monochromatic wave. Such a one would have, theoretically, started an eon ago and we would have to wait for the millennium to see it. Practically, a quasi-monochromatic wave has a very narrow spectral band and very little energy. Thus, every radiometric measurement is made over a finite spectral band. This means that the voltage V from a source $S(\lambda)$ that is detected by a detector with a spectral response $\Re(\lambda)$ is given by

$$V = \int S(\lambda)\Re(\lambda)d\lambda. \qquad (9\text{-}1)$$

Under only very restricted conditions can the flux in the band be measured. It is useful to write this another way. The output voltage (current, charge, or other) is given by

$$V = S\Re \int s(\lambda)r(\lambda)d\lambda. \qquad (9\text{-}2)$$

This formulation emphasizes that the source is characterized by a relative spectral distribution, $s(\lambda)$, with maximum value of 1 and a constant, S, that "de-relativizes" it. The same is true for the detector responsivity, $\Re r(\lambda)$. Thus, based on Equation 9-2, no matter how clever you are, no matter what you do, unless you have a flat detector, the flux in the band cannot be measured. Only the flux weighted by the detector response can be measured. If the detector has a flat response, i.e., one that does not change with wavelength, then $r(\lambda) = r = 1$, and

$$V = S\Re \int s(\lambda)d\lambda .$$ (9-3)

However, such a flat detector is an experimentalist's dream; it can only be approximated, and it is often not very sensitive when it is.

In the real world, a fairly wide bandwidth is often used, and the flux that is measured is the flux that is weighted by the spectral response of the detector, a so-called *effective* value.

9.2 EFFECTIVE VALUES

Before going further, a word of warning. When a colleague uses *effective* to describe something, stop. Ask for meaning. Be skeptical. An *effective* value can mean almost anything. It can mean very efficient, as in, "I introduced a very effective, new procedure". It can mean it wasn't really so, as in, "This house (or car) is effectively brand new." An effective value can have a definite meaning (that may not be shared by all). It is the value of a quantity that effects a response from a receiver. An effective watt can be written as an equation:

$$\Phi_{eff} = \int \Phi(\lambda)\Re(\lambda)d\lambda .$$ (9-4)

This represent the flux or power as it causes a response in a given detector. Clearly (I think) there are also *effective* radiances, exitances, incidances, and intensities. They are various geometric fluxes weighted by a detector response. The best-known effective watt is the power that causes a response in the human eye, and that gives (and gave) rise to the venerable field of photometry.

9.3 PHOTOMETRY

The important weighting function in photometry is the response of the human eye. It is, in fact, the response of the average human eye under reasonably bright lighting conditions, called the photopic response. Data are available from a host of measurements, but first consider the calculations. The response of the human eye is given special symbols, so that the response is

$$\Re_{eye} = K_m V(\lambda).$$ (9-5)

Then, the lumen, which is the eyeball effective watt, is given by

$$\Phi_v = K_m \int \Phi(\lambda) V(\lambda) d\lambda \,,$$ (9-6)

where K_m is 683 lm W^{-1} at 540 tHz and $V(\lambda)$ is in Figure 9-1 and Table 9-1.

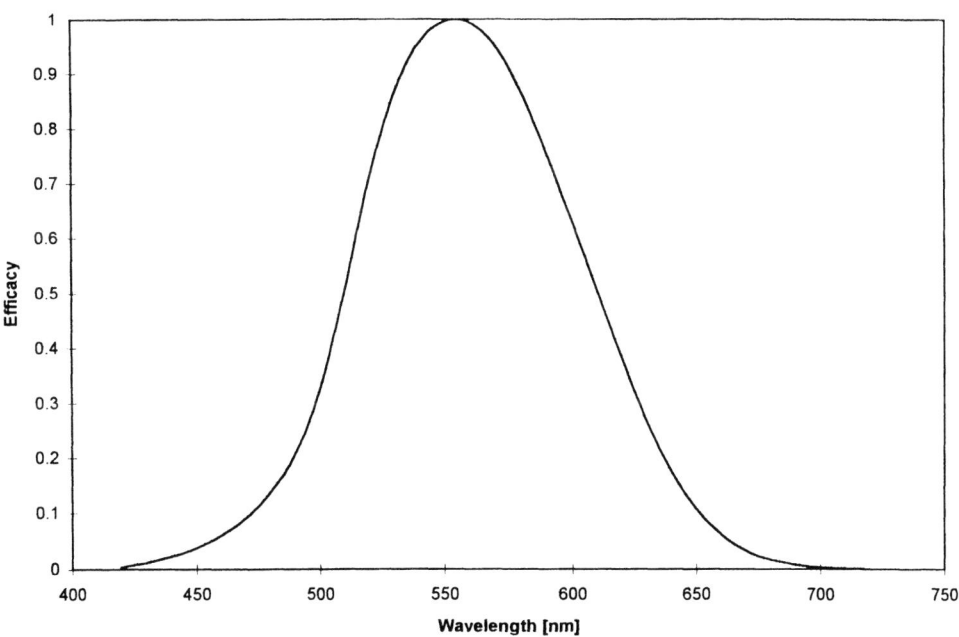

Figure 9-1. Luminous efficacy.

Table 9-1. The visibility factor.			
Wavelength [nm]	$V(\lambda)$	Wavelength [nm]	$V(\lambda)$
400	0.0004	590	0.757
410	0.0012	600	0.631
420	0.0040	610	0.503
430	0.0116	620	0.381
440	0.023	630	0.265
450	0.038	640	0.175
460	0.060	650	0.107
470	0.091	660	0.061
480	0.139	670	0.032
490	0.208	680	0.017
500	0.323	690	0.0082
510	0.503	700	0.0041
520	0.71	710	0.0021
530	0.862	720	0.00105
540	0.954	730	0.00052
550	0.995	740	0.00025
560	0.995	750	0.00012
570	0.952	760	0.00006
580	0.87		

The rest of the photometric quantities derive from this special effective power. Table 9-2 gives the basic terms and units in photometry with corresponding radiometric terms.

General Name	Photometric		Radiometric
	Name	Units	Units
Energy	Luminous Energy	talbot	J
Energy Density		talbot m^{-3}	J m^{-3}
Flux	Lumen	lm	W
Exitance	Luminous Exitance	lm m^{-2}	W m^{-2}
Incidance	Illuminance	lm m^{-2}	W m^{-2}
Intensity	Luminous Intensity	lm sr^{-1} candela (cd)	W sr^{-1}
Radiance	Luminance	lm m^{-2} sr^{-1} cd m^{-2}	W m^{-2} sr^{-1}
Fluence	Exposure	lm s	J

Table 9-2. Photometric symbols and units.

The lumen per square meter is also called the lux, and a millilux is a nox. The English variety is the lumen per square foot, also called a foot-candle. Similarly, the lux is also called a meter candle. The intensity unit is the candela, and is the same as the lumen per steradian. Luminance can be given as candela per square meter or candela per square centimeter, which is also called a stilb. An apostilb is a stilb divided by π, and a foot lambert is a foot candle divided by π. A skot is a milliapostilb. No wonder some say that the worst part of photometry is the units!

Although this text is not about photometry, some notice may be taken as to how to approach this Tower of Babel. Understand the geometric concepts, as described in the earlier chapters. Look up the units in proper tables. Convert them numerically to a set that resembles those above, and proceed as with radiometry. Tables 9-3 and 9-4 give many of the units and conversions that should help.

Table 9-3. Luminance units and conversions.

	cd m^{-2} nit	cd cm^{-2} stilb	cd ft^{-2}	cd in^{-2}	apostilb blondel	milli lambert	foot-lambert
cd m^{-2} =	1	10,000	10.764	1550	1/π	10/π	3.426
cd cm^{-2} (stilb) =	0.0001	1	0.001076	0.155	3.183×10^{-5}	π/10,000	0.0003426
cd ft^{-2} =	0.0929	929	1	144	0.02957	0.2957	π/10
cd in^{-2} =	0.000645	6.452	0.00694	1	0.0002054	0.002054	0.002211
apostilb =	π	10,000π	33.82	4869	1	10	10.764
milli lambert =	0.1π	1,000π	3.382	486.9	0.1	1	1.0764
foot lambert =	0.2919	2919	π	452.4	0.0929	0.929	1

	ft-cd	lm m⁻² (lux)	phot	milliphot
Table 9-4. Illuminance conversions.				
ft-cd =	1	0.0929	929	0.929
lm m⁻² =	10.764	1	10,000	10
phot =	0.00108	0.001	1	0.001
milliphot =	1.076	0.1	1,000	1

9.4 AN ILLUMINATING EXAMPLE

This example is another *twofer* or *threefer* or *morefer*. That is, it illustrates several different points. It tells the conditions under which normalization does (and does not) apply. It tells how one can use normalization correctly. It shows the conditions for luminous calculation and for calculation with other normalization schemes.

The example is the detection of a star that provides an illuminance at the surface of the Earth of 1 microlumen per square meter. A photomultiplier with a responsivity of 0.01 amps per lumen will be used to detect a star that is similar to our very own sun, which has a temperature of 6000 K. A telescope with an aperture area of 1 square meter will also be used to collect the light.

The solution to the problem seems easy. Multiply 1 microlumen per square meter by 1 square meter and by 0.01 amps per lumen and get 0.01 microamps for the output. That would be wrong. The star does not have a response that matches the eye response, and neither does the responsivity of the tube.

The correct solution, to be described immediately, can also be generalized to other normalizations. The only generally correct way to obtain an output, say in amps, is to integrate the product of the spectral distribution of the source and the spectral responsivity of the sensor:

$$I = \int \Re(\lambda) S(\lambda) d\lambda, \tag{9-7}$$

where $S(\lambda)$ is the spectral distribution of the source and $\Re(\lambda)$ is the spectral responsivity of the detector, as before. The photomultiplier tube is calibrated by shining a calibration source on it. The output current is given by

$$I_c = \int \Re(\lambda)S_c(\lambda)d\lambda = \Re S_c \int r(\lambda)s_c(\lambda)d\lambda, \qquad (9\text{-}8)$$

using the capital and lower case conventions wherein capitals are numerical values and the lower case distributions are normalized to one. There are no subscripts on the absolute and spectral responsivities since they are invariant for a stable instrument over the calibration and measurement. The luminous responsivity of the tube is given by

$$\Re_v = \frac{\Re S_c \int rs_c\,d\lambda}{VS_c \int s_c v d\lambda}, \qquad (9\text{-}9)$$

where the wavelength dependence of the lower-case distributions has been suppressed. The numerator of the right-hand side is the output one gets when the source is shone on the detector. The denominator represents the number of lumens that are in the source. Equation 9-9 is the responsivity in amps per lumen. The visibility curve is written as $Vv(\lambda)$, and as Vv when the wavelength variable is repressed. Note that the luminous responsivity does have a subscript because it is a special type. This is the calibration, and the constant of the calibration is given by rearrangement of Eq. 9-9:

$$\Re = \frac{\Re_v V \int vs_c\,d\lambda}{\int rs_c\,d\lambda}. \qquad (9\text{-}10)$$

Then it is possible to measure the star. These measurement quantities are not subscripted. The actual current that is measured is given by

$$I = \Re S \int rs d\lambda. \qquad (9\text{-}11)$$

The luminous flux from the star is given by

$$S_v = SV \int svd\lambda, \qquad (9\text{-}12)$$

and

$$S=\frac{S_v}{V\int svd\lambda}.$$ (9-13)

The output current can then be written

$$I=\Re S\int rsd\lambda=\left[\frac{\Re_v V\int s_c vd\lambda}{\int rs_c d\lambda}\right]\left[\frac{S_v}{V\int svd\lambda}\right]\int rsd\lambda,$$ (9-14)

and this can be simplified to

$$I=\Re_v S_v\left[\frac{\int s_c vd\lambda}{\int svd\lambda}\frac{\int rsd\lambda}{\int rs_c d\lambda}\right].$$ (9-15)

The output current is equal to the luminous responsivity times the luminous flux *only if* the bracket as a whole goes to one. This can happen if the measuring detector has the same relative spectral responsivity as the eye, i.e., $r = v$, or the calibration source has the same relative spectral output as the unknown, i.e., $s_c = s$. Although this development has been carried out for luminous quantities, it applies equally well to any other normalization. In the general case, Vv can be replaced by the absolute spectral response of the detector in question.

9.5 OTHER NORMALIZATIONS

Normalization schemes are important conveniences. Photometric instruments are made with appropriate filters that have the eye response for making the measurements in order to satisfy the requirements delineated by Eq. 9-14. A different normalization scheme was used by those developing the Sidewinder guided missile. All measurements were made with lead sulfide detectors and normalized to that response because the guidance unit would use a lead sulfide detector. The output was measured in "lead sulfide watts."

Photosynthetic processes occur at about 0.8 μm in various plants. Because the spectral band has a finite width, this entire operation can be normalized to the "photosynthetic watt."

Sunburn is caused by radiation at about 0.3 μm. This normalization scheme is the "sunburn watt" and even has a name like the lumen. The sunburn watt is the finsen.

9.6 NORMALIZATION TO THE PEAK

The normalization schemes discussed above are all normalizations to spectral distributions. It is better not to have to normalize, but conditions often require it. Recall that a radiometer provides an output voltage (or current or digital count) that is the integral of the product of the absolute spectral flux from the source and absolute spectral response of the detector,

$$V = \int S(\lambda) \Re(\lambda) dl \; , \qquad (9\text{-}16)$$

where the integration is from the minimum wavelength of the detector response to its maximum. This is illustrated in Figure 9-2, where both the responsivity and the spectra are shown. The spectra are those of two blackbody distributions, one from a 300 K temperature; the other for 3000 K. The responsivity is a gaussian. The magnitudes of the two spectra were chosen so that the two integrations give the same value. How does one deal with this?

One approach is to write the normalized flux Φ_n, as

$$\Phi_n = V / \Re_n . \qquad (9\text{-}17)$$

The issue, then, is how to define the normalized responsivity. The most common way to do this is normalization to the peak, i.e.,

$$\Re_n = \Re(\lambda_m) \; , \qquad (9\text{-}18)$$

where $\Re_n$ is the response at the peak. Then the product of the responsivity at the peak and the flux there must be equal to the integral of the product:

$$V = \Re_p S_p = \Re S \int rsd\lambda . \qquad (9\text{-}19)$$

Then, when one measures an output electrical signal, he can refer to it as a peak normalized value:

$$V_{np} = \frac{V}{\mathfrak{R}_p}.$$

(9-20)

Other terms are equivalent radiation at the peak and peak normalized radiation at λ_p or λ_m. It is important when using this shorthand that the real measurements are kept in mind.

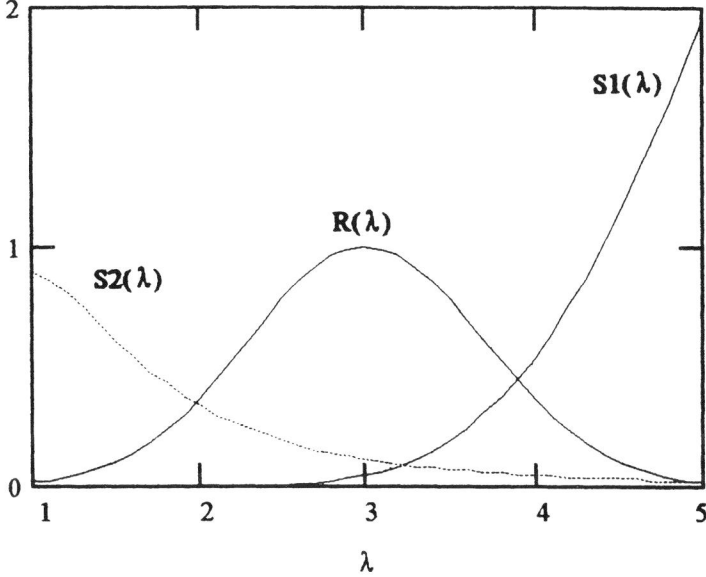

Figure 9-2. Illustration of two different sources providing identical outputs. Both S1 and S2 are blackbody distributions, S1 at 300 K, the other at 3000 K.

9.7 NORMALIZATION TO THE AVERAGE

If there is no well-defined peak, normalization to the average is appropriate. Then the output voltage can be written

$$v = \overline{\mathfrak{R}}\,\overline{S} = \mathfrak{R}S \int rs d\lambda.$$

(9-21)

The same caveats about this shorthand are applicable.

9.8 NORMALIZATION TO THE BANDWIDTH

Although normalization to the bandwidth is used in the usual, optical spectral sense, it is more often used in terms of an effective noise bandwidth in which the spectrum is a temporal one. The total noise of a system is the square root of the integral of the square of the noise spectrum. In practice the bandwidth of a system is always limited somehow, and it is often set by the required information bandwidth. Then the bandwidth normalization involves setting a certain bandwidth and a constant noise level such that their product yields the appropriate value that is really obtained by the integral of the noise spectrum:

$$V_n^2 = \int_{\Delta f} V^2(f)df = V_0^2 \Delta f. \qquad (9\text{-}22)$$

One may think of the bandwidth-normalized noise as the noise in a rectangle with zero as the base, the rms noise level as the top, and the upper and lower frequencies as the sides. The area of this rectangle must be equal to the area under the curve of the rms noise spectrum.

9.9 A NASTY DENORMALIZATION

This example illustrates how one can (must) unfold some normalizations to get a correct answer. The problem was—will a night vision photomultiplier system work at one-quarter moon? As usual, the correct expression is

$$V = \int S(\lambda)\Re(\lambda)d\lambda . \qquad (9\text{-}23)$$

So we need both the absolute spectral responsivity of the tube and the absolute spectral irradiance on the detector. The tube data are published, and can be assumed known. However, the best data on the quarter moon are given in foot candles. We convert these to lumens per square meter using Table 9-4. We also realize that, to a very good approximation, moonlight has the same spectral response as sunlight, essentially that of a 5900 K blackbody. Thus,

$$E_v = S \int s(\lambda)d\lambda \qquad (9\text{-}24)$$

and

$$S = \frac{\int s(\lambda)d\lambda}{E_v} . \tag{9-25}$$

Both terms on the right are known, so S is known, and therefore $S(\lambda) = Ss(\lambda)$ is known. It is relatively easy to write these equations, but there is much number crunching required to evaluate this.

9.9 RECAP

No real, physical quantity can be measured in an infinitely narrow bandwidth. The output is therefore the product of the input spectral flux and the spectral responsivity. There is no unique value for the integral of this product. It can be done in the special cases for which the responsivity is flat or both the spectra are known. Several normalization schemes have been developed to avoid making spectral radiometric measurements. The most notable is photometry. It was shown here that one can obtain the true output by multiplying the luminous input by the luminous response *if* either the calibration source and the measured object have the same relative spectrum or the radiometer has a luminous responsivity. One can normalize to the peak, average, bandwidth, or even arbitrarily, but no normalized quantity can be relied on as equaling the integral of the spectral flux.

CHAPTER 10

CALIBRATION STANDARDS

Most radiometric measurements require that calibration by the use of standards be performed prior to the measurements of unknowns. The requirements for standards fall naturally into two different types: flux measurements and material measurements. The flux measurements include power, incidance, and radiance, while the material standards include those for reflectance, transmittance, emissivity, absorptance, detector responsivity, and refractive index.

10.1 TYPES OF STANDARDS

Standards may be divided into primary and secondary standards as well as flux and material standards. Another type, the need for which has been created more by lawyers and contract managers, is the traceable standard. This is one that can be traced to an appropriate official agency for purposes of satisfying a contract.

A primary standard is based on the measurements of quantities other than that of the standard. For instance, one primary standard is a cavity radiator at the National Institute of Science and Technology (NIST) formerly the National Bureau of Standards (NBS). This cavity radiator has its emissivity calculated by one of the methods described above to be at least 0.9999, and its temperature is then measured. By measurement or calculation of two non-flux quantities (temperature and emissivity), the flux from the cavity is determined. Such is a primary standard, and only NIST, or other national standards labs, have such standards. NIST also has secondary (or working) standards. One example is a set of tungsten bulbs that are compared to the primary cavity radiator just described, and are used in other laboratories as calibration sources. It is often good practice in individual labs to follow this procedure of using a primary and several working standards. In this case, the "primary" standard is actually a secondary standard obtained from NIST, and the working standards are compared to it as often as is necessary.

10.2 PHOTOMETRIC STANDARDS

NIST has been required by Congressional mandate to maintain the standard of luminous intensity, a long, cylindrical body with a conical end. The tube, shown in Figure 10-1, is made of platinum and is surrounded by an insulator made of unfused thoria and then fused thoria. The temperature is monitored in several places with thermocouples (and sometimes other temperature sensors), and the emissivity was calculated by the methods of Gouffé and DeVos. The value quoted is 0.9999+ as a result of its very long aspect ratio that induces many, many reflections and therefore absorptions.   The definition of the candela (the unit of luminous intensity) up until 1979 was the intensity from 1/60th square centimeter of a blackbody at the temperature of fusing platinum.   Therefore, the cylinder was surrounded by pure platinum that was heated above its melting point.   The time-temperature curve is monitored as the platinum is heated.   First, the temperature increases until the melting point is reached.   Then it flattens as the heat is taken up in the heat of melting until it rises again when all the metal has been melted. Because the definition called for the temperature of the fusing platinum, the cooling curve is used.   The temperature comes back down until it flattens as the heat is taken up in fusion, and that is the magic moment—or really about five minutes.

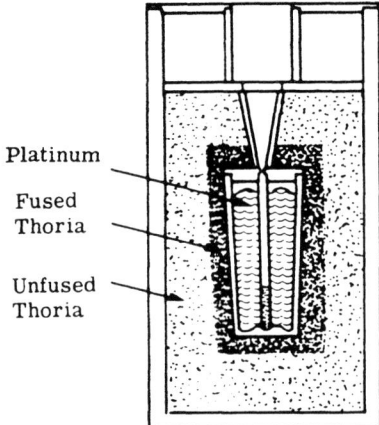

Platinum

Fused
Thoria

Unfused
Thoria

Figure 10.1. The luminous standard.

The errors have been analyzed; they arise from impurities in and the thermodynamic freezing point of platinum, but not in the inaccuracies of the thermometers.   There is a very small uncertainty in the calculated emissivity because the cavity has been overdesigned; it is very long and narrow.

The candela was redefined by the General Conference on Weights and Measures as follows:

> The candela is the luminous intensity, in a given direction, of a source that emits monochromatic radiation of frequency 540×10^{12} Hertz and has a radiant intensity in that direction of 1/683 watts per steradian.

> The candela so defined is the base unit applicable to photopic quantities, scoptopic quantities and to be defined in the mesopic range.

(Photopic quantities mean, essentially, the eye operating with normal illumination. Scotopic and mesopic regions for which there is less light, were not considered.) Therefore by definition K_m is 683 lumens per watt. No new standard has been introduced based on this new definition. In fact, it is not necessary.

The cavity was used most recently in 1931 to calibrate a bank of tungsten bulbs, and they in turn have been used since to calibrate others. Although these are secondary standards, they have been used as primary standards all this time. Other standards organizations in Australia, Canada, France, Germany, and Great Britain have also developed luminous standards. The International Bureau of Weights and Measures (BIPM, Bureau de Poids et Measures) in Sevres, France, made an intercomparison. They prepared two different types of gas-filled incandescent lamps (Osram Wi41G and NPLGEC) and sent them to the laboratories. Each of the laboratories measured either four or six of either type that were operated at a color temperature of 2800 K under constant-current conditions with a fixed polarity. (Color temperature is defined and discussed in Chapter 14 on radiation temperatures.) It was a radial-type round robin. The lamps were sent to each laboratory by BIPM, returned to BIPM, and sent to the next lab. Both intensity, the candela, and power, the lumen, were measured. Table 10-1 summarizes the results. The entry BIPM/others indicates value measured by BIPM over the average of the others. The standard deviation of these is also shown, while the final line shows the ratio of BIPM to NBS—when it was NBS.

Table 10-1. Photometric standard intercomparisons.		
	Candela	Lumen
Number of Labs	15	11
BIPM/Others	0.990	1.007
Standard deviation	0.77	0.58
BIPM/NBS	0.985	0.997

This indicates, but does not prove, that the NIST uncertainty is 1.5 percent for luminous intensity and 0.3 percent for power, but it really just means that the different labs are different by this much. A simple thing to remember is that the various national standards labs have a disagreement of only about 1%. If only our diplomatic and political disagreements were this small.

NIST performs three types of photometric calibration: luminous intensity (the candela), luminous flux (the lumen), and color temperature (kelvins). The process is to use the gold-point blackbody to calibrate a variable-temperature blackbody, which is then used to establish scales of spectral radiance and spectral irradiance. From these the three photometric quantities are obtained for the type of lamp desired by the customer[1].

10.3 FLUX STANDARDS

These standards come in two varieties, source standards and receiver standards. The NIST standards have classically been a set of tungsten bulbs calibrated against a cavity radiator. NIST also has a gold-point cavity radiator. They have developed a cavity receiver, but it has not yet been accepted as an official standard. The same is true of a newly developed detector standard based on a calibrated silicon detector. Several other national labs have so-called active cavity standards. All of these are discussed in this section.

10.3.1 SOURCE STANDARDS

These are all based on the same principles as the standard of luminosity. A cavity is formed that is usually a right circular cylinder truncated with a cone. It is made long enough that the emissivity of the hole is at least 0.9999. The cavity is surrounded by the working material. In the case of the luminous standard it is platinum; for an infrared standard it is gold. The working material is as pure as possible and is heated to its melting point. Depending on the definition, it is then used at the melting point or at the fusing point by employing the time-temperature curve discussed above.

Probably the first use of a cavity radiator as a standard was that of Coblentz of the NBS in 1915[2]. He used a ceramic cavity in the range of 1300 K to calibrate carbon filament lamps as secondary standards. These standard lamps were in use until the late 1940s in the USA and elsewhere. The standard tungsten lamps that

---

[1]Booker, R. L. and D. A. McSparron, *Photometric Calibrations*, NBS Special Publication 250-15, U.S. Department of Commerce, 1987.

[2]Coblentz, W. W., "The present status of the standards of thermal radiation maintained by the National Bureau of Standards," Bulletin of the National Bureau of Standards **11**, 87 1915.

are in use today are very similar. In 1960[3] Bedford of the Canadian Research Council used a large, conical cavity at about 425 K, and there was a radiation difference between the two that corresponded to a temperature difference of 0.3 K.

The gold-point standard[4] is shown schematically in Figure 10-2. It is useful in the spectral region from 1.5 to 14μm with an aperture no larger than 2×4 mm. The standard is used for checking other sources; it cannot be purchased; it has an uncertainty of about 2%; it is a traceable standard.

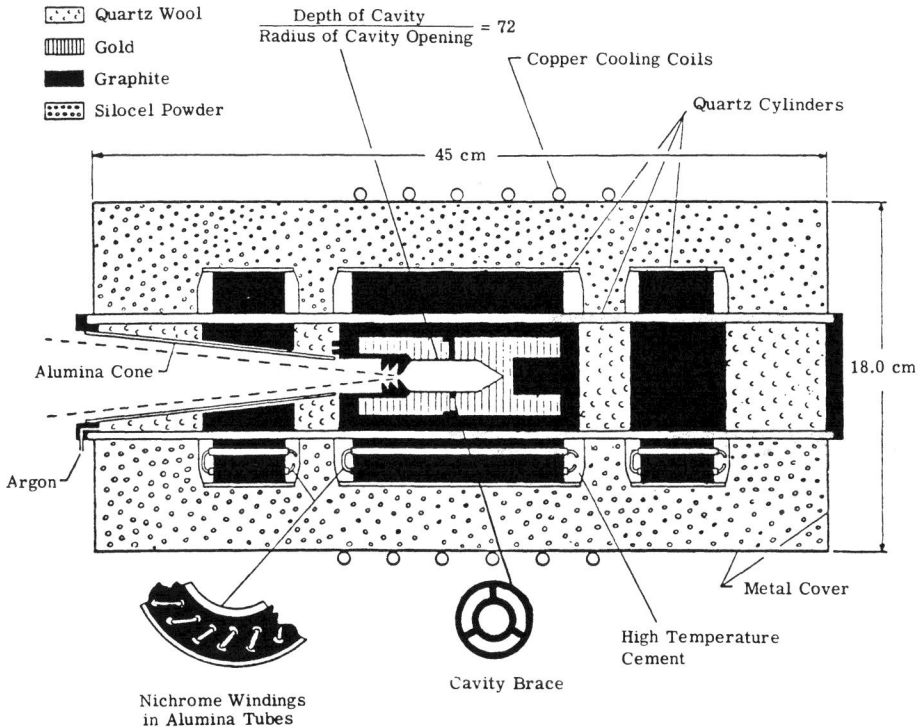

Figure 10-2. The gold-point blackbody.

---

[3]Bedford, R. E., "A low temperature standard of total radiation," Canadian Journal of Physics **38**, 1256 1960.

[4]Lee, R. D., "The NBS photoelectric pyrometer and its use in realizing the International Practical Temperature Scale above 1063°C," Metrologia **2**, 150, 1966.

The tungsten lamps have been cross-calibrated and can be purchased from companies such as Eppley and Optronic. They have a three-sigma uncertainty ranging from about 1% in the middle of the visible to about 2.5% at 0.25 μm and 2.5 μm.

The calibration process starts with the use of the gold-point blackbody, the variable temperature blackbody mentioned above, and a set of lamps. The gold-point body is the primary standard; the variable-temperature blackbody and the lamps are the secondaries. The FEL and other lamps are calibrated, and the information is sent to the user. The same process is used for both radiance[5] and irradiance[6] calibration of lamps and is almost identical to the photometric calibration process.

10.3.2 ELECTRICAL SUBSTITUTION RADIOMETERS

These devices, abbreviated ESRs and called alternately electrically calibrated radiometers, ECRs, and absolute radiometers, are essentially cavities used as receivers rather than sources. The basic concept is to use a cavity to collect radiation from a source and read the electrical output from it. Then the cavity is heated by electrical wires until the same output is found. The amount of joulean heating required from the electrical circuit determines the power of the source, since they are equal. One version, developed by JPL, is called an active-cavity radiometer, or ACR.

Blevin and Brown[7] in 1971 used one of these devices to measure the Stefan-Boltzmann constant. Theirs was the first measurement that agreed with the theoretical value within the experimental error. The value based on atomic constants is $5.6705\pm0.00019\times10^8$ W m^{-2} K^{-4}, so that the error is 0.00335%. Their uncertainty was about 0.1%.

---

[5]Walker, J. H., R. D. Saunders, and A. T. Hattenburg, *Spectral Radiance Calibrations*, NBS special Publication 250-1, U.S. Department of Commerce, 1987.

[6]Walker, J. H., R. D. Saunders, J. K. Jackson, and D. A. McSparron, *Spectral Irradiance Calibrations*, NBS Special Publication 250-20, U.S. Department of Commerce, 1987.

[7]Blevin, W. R. and W. J. Brown, "A precise measurement of the Stefan Boltzmann constant," Metrologia **7**, 15, 1971.

More recently Quinn and Martin[8] of Great Britain's National Physical Laboratory (NPL) have measured the Stefan-Boltzmann constant with an order of magnitude greater accuracy (and difficulty). They used the triple point of water for the source temperature and operated their ESR at liquid helium temperatures! This puts the uncertainty at about the same level as that of the atomic constants (0.00335% vs. 0.01%). This is probably the most accurate radiometer ever made and used.

Geist and Blevin[9] have also built and used an ESR for NIST, and Boivin and McNeely[10] of the National Research Council (NRC) have done it in Canada. JPL had made several for space simulators and measurements; the World Radiation Center (WRC) in Davos has made some for measuring the solar constant[11]; and Cambridge Research Instruments (CRI) makes at least two different varieties that are for sale[12]. The properties of these and other standards are summarized in Table 10-2.

---

[8]Quinn, T. J. and J. E. Martin, "Radiometric measurements of the Stefan Boltzmann constant and thermodynamic temperatures between -40°C and 100°C," Metrologia **20**, 163, 1984.

[9]Geist, J. and W. R. Blevin, "Chopper stabilized null radiometer based upon an electrically calibrated detector," Applied Optics **12**, 2532, 1973.

[10]Boivin, L. and F. T. McNeely, "Electrically calibrated absolute radiometer suitable for measurement automation," Applied Optics **25**, 554, 1986.

[11]Kendall, J. M. and C. M. Berdahl, "Two blackbody radiometers of high accuracy," Applied Optics **9**, 1082, 1970.

[12]Cambridge Research Instruments Company, Cambridge, MA.

	Units	NBS	NPL	NRC	WRC	CRI
Size	mm		40×200		35×50	
Aperture	mm		12		5	
Responsivity	VW^{-1}	0.034	0.07	0.09	0.04	
Time constant	s	10	180	15	1.7	
NEP		100nW	2nW	50nW		
Uncertainty	%	0.18	0.02	10	0.22	0.04
Spectrum	μm	total	total	0.2-3.0	total	total

Table 10-2. Electrical substitution radiometer characteristics.

10.3.3 SELF-CALIBRATING DETECTORS

One of the most convenient and accurate standards yet developed is the so-called self-calibrating detector, first proposed by Geist[13]. In its present form it is a silicon *p-n* junction, so that it can only be used in the spectral region from 0.4 to 0.95 μm. The reflectivity of the detector is measured by one of the standard techniques discussed later. Silicon is sufficiently specular that the scattered component of reflection is insignificant. Silicon also has a very high absorption in this spectral region so that the internal quantum efficiency is very high. It is made to saturate by increasing the back bias voltage. When these measures have been taken, the external quantum efficiency, i.e., the photon responsivity, is known. If the responsivity is known well enough, then the device is a detector standard.

How do you check to see if a primary standard is accurate? You can only compare. If you had another, more accurate standard, why bother with this one? Comparison is what Geist and Zalewski did. Table 10-3 shows this. They measured a laser with two photodiodes that had been appropriately characterized and with an ESR. The table shows the power in the beam as measured by the two diodes, SN11 and SN190, the ESR, and the ratios.

---

[13]Geist, J., "Quantum efficiency of the p-n junction in silicon as an absolute radiometric standard," Applied Optics **18**, 760, 1979.

Table 10-3. Comparisons of detector standards.				
Run	SN11	SN190	ACR	SN11/SN190	SN/ACR
1	1.0117	1.0114	1.0096	1.0030	1.0019
2	1.0131	1.0131	1.00116	1.0000	1.0015
3	1.0132	1.0133	1.0148	0.9999	0.9985

Drawing statistical inferences from so little data is dangerous, but several things are apparent. The run-to-run variability is clear: run 1 was different. Perhaps the laser was unstable, or the moon was blue. Considering just the last two runs, the two detectors differed by only 0.0001, and they differed from the ESR by only 0.0015. The ESR has an estimated uncertainty of 0.001. The self-calibrating detector appears to have an uncertainty of about 0.1% and is extremely easy to use and build. It is also quite sensitive, but it is severely limited in the spectral range that it can cover.

There are commercial versions of the self-calibrating detectors, and they generally come in an improved geometry that eliminates the need for measuring the reflectivity. Such a trap detector is shown in Figure 10-3.

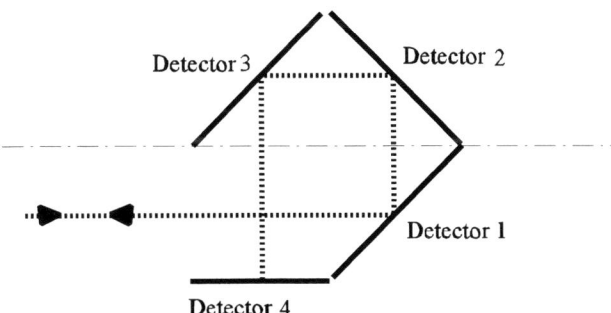

Figure 10-3. The trap detector. Light enters and is partially absorbed by detector 1. The rest is reflected to detector 2, which repeats the process. The light undergoes seven reflections before it exits. The detector outputs are added.

The fact that there are seven reflections (and absorptions) at about 0.1 means that the total reflection is 0.1×10^{-6}; all of the radiation is absorbed!

10.4 SYNCHROTRONS

These devices have been proposed as standard radiators for the ultraviolet and shorter wavelengths. They will not be discussed further in this text[14].

10.5 MATERIAL STANDARDS

These are materials whose transmittance and reflectance have been carefully calibrated, and can therefore be used in the laboratory as references and with traceability.

10.5.1 REFLECTANCE STANDARDS[15]

Several material standards are available from NIST. There are two first-surface mirror standards made of gold and aluminum 5.1 cm in diameter and a second surface aluminum mirror that is 5.1 cm square. These are used for specular reflectivity standards.

There are four diffuse standards: opal glasses of two different sizes for the spectrum from 400 nm to 750 nm, white ceramic tile for 350 nm to 2500 nm, and black porcelain enamel tile for 280 nm to 2500 nm.

There are four wavelength standards, one to be used in reflectance, the others in transmittance. Holmium oxide is good for 240 nm to 640 nm; a rare-earth mixture is used for 740 nm to 2000 nm; Didymium glasses are used for 400 nm to 750 nm (one for reflectance; the other for transmittance).

---

[14]Riehle, R. and B. Wendy, "Electron storage ring BESSY as a radiometric source of calculable spectral radiant power between 0.5 and 1000 nm," Optics Letters **10**, 365, 1985.

[15]Weidner, V. R. and J. J. Hsia, *Spectral reflectance*, NBS special Publication 250-8, U.S. Department of Commerce, 1987.

10.5.2 Transmittance Standards[16]

Eleven different types of materials are available, including quartz cuvettes, potassium dichromate, glass filters, and liquids. Seven measurement services are available, all variations of the measurement of spectral transmittance of samples sent by users. NIST needs to be contacted for applicability, sizes, costs, and duration of tests.

10.6 Recap

The most frequently used standard is the tungsten bulb that emits nearly graybody (the emissivity decreases from about 0.4 to 0.2 from 0.4 μm to 2.5 μm) radiation at a temperature of about 2800 K and can be purchased for individual laboratory use. There are also gold-point cavity radiators that can be calibrated by NIST and then used in the individual laboratory. Absolute radiometers exist in many standard laboratories, principally to calibrate sources and for research. Silicon detectors can be characterized to serve as receiver standards. Table 10-4 summarizes their pertinent properties.

Table 10-4. Summary of standards.				
Type	Uncertainty [%]	Spectrum [μm]	Level [W]	Cost [$]
Tungsten	1-2.5	0.25-2.5		200
Gold Cavity	5	1-20		20,000
ESR	0.01-0.1	1-20	10^{-9}	10,000
SCD	0.1	0.5-0.95	10^{-11}	500

The cost numbers are my estimates. They are not vendor quotes. The only vendors I know for the gold-point cavity are EOS and Mikron, and Cambridge and

---

[16]Eckerle, K. L., J. J. Hsia, K. D. Mielenz, and V. R. Weidner, *Regular spectral transmittance*, NBS special Publication 250-6, U.S. Department of Commerce, 1987.

Oxford for the cryogenic ESR.

Catalogs of services and standards supplied by NIST are available. The most recent is the 1998 version[17].

---

[17]Standard Reference Materials Catalog, U.S. Department of Commerce and NIST Calibration Services Users Guide. By phone: 301-975-6776 and 301-975-2002, respectively.

CHAPTER 11

MEASUREMENT TECHNIQUES

There are many different types of radiometric measurements. They may be classified as measurements of flux in its many different geometric forms; measurements of the material properties of transmittance, reflectance, emissivity, absorptance, and scattering; and properties of instruments. Other aspects of measurement include the restrictions and effects of the different independent variables: spectrum, time variation, angle, position, and extent. Approaches to making the measurements and accounting for the extraneous and intrinsic influences are given, and some of the applications are contained in this chapter.

11.1 RELATIVE AND ABSOLUTE MEASUREMENTS

Absolute measurements are always harder to make than relative ones, like absolute truth is harder than relative truth (honor and integrity)! A relative measurement may be considered to be the ratio of output signal levels. It may be a set of signals compared to the first one, as in a relative spectral transmission. The value of the transmission is never determined, only how it changes from one part of the spectrum to another. An absolute measurement requires a calibration so that the value of the radiometric quantity can be found. For this example it would be the transmittance of the sample, perhaps at all wavelengths of interest. Sometimes these can be combined: an absolute measurement is made at one wavelength, and relative measurements are made with respect to this value.

11.2 ERRORS

The quality of a measurement is usually described in terms of its accuracy and precision. Actually, most people state the *inaccuracy* and *imprecision* since low numbers are cited. Typically one might cite an accuracy of 1%, actually meaning an uncertainty in the accuracy of 1%, but that is tradition and shortspeak.

Accuracy is the degree to which the actual value is measured. The uncertainty in accuracy is a value that represents how close to truth the measurement probably comes. Precision is somewhat easier to grasp. It is the repeatability of a measurement. The uncertainty in accuracy can be no less than the variability, the lack of repeatability, the (im)precision of the measurement.

In a sense, absolute measurements are made relative to a standard, one of the radiometric standards or one of the material standards described in the previous chapter. So everything may be relative.

Error analysis is an extensive subject. Only a few of the basic concepts are used here. In particular, use is made of the standard technique of evaluating an error based on a describing equation. To obtain the relative error of a given expression, the differential of the logarithm of the expression is taken. For instance, if a certain function $f(x)$ is given by

$$f(x) = ax - bx^2 + \frac{1}{x} \; , \qquad\qquad (11\text{-}1)$$

then the relative error is

$$RE = \frac{d}{dx}\{[\ln[f(x)]]\}dx = \frac{d}{dx}[\ln(a) + \ln(x) - \ln(b) - 2\ln(x) - \ln(x)]dx$$

$$= 0 + \frac{dx}{x} - 0 - 2\frac{dx}{x} - \frac{dx}{x} \; . $$

$$(11\text{-}2)$$

This makes sense. The relative change in $f(x)$ is the change divided by the amount. If the function is squared, the error is twice as large, and if it is negative, the error is in the opposite direction. Finally, if it is a reciprocal, it is also in the opposite direction. If the three terms are independent, then they must be added in quadrature: each term is evaluated independently and the root-sum square is taken. The result for this example is

$$RE = \sqrt{\left[\frac{dx}{x}\right]^2 + \left[2\frac{dx}{x}\right]^2 + \left[\frac{dx}{x}\right]^2} \qquad\qquad (11\text{-}3)$$

Since they are independent, each contributes an additive error, even though they may be in opposite directions.

11.3 RULES OF MEASUREMENT

Every measurement requires a calibration before, during, or after the measurement, or all three. There seem to be only two completely general rules for making radiometric measurements:

Calibrate like you measure.
Think of everything.

Some things to consider, based on the first rule are: if a point source is to be measured, use a point source for calibration, match the speed of the optical systems used for both; use the same spectral range, same temperatures, degree of coherence, spatial extent, field of view, and any other similarities that make sense.

A somewhat *extreme* example consists of the measurement of the radiation from the planet Mars with an instrument here on the surface of the Earth. The best calibration that can be obtained is to put a standard source right next to Mars with the same angular extent as Mars, the same spectral distribution, and the same radiant exitance. Then the calibration is just like the measurement. Of course, if all that were known, why make the measurement?

The second rule may be implemented (at least in part) by systematic procedure of listing every independent variable in the responsivity of the radiometer or other instrument and the measurement experiment.

11.4 THE MEASUREMENT EQUATION

Most radiometric measurements are made with a radiometer, and even those that do not use one can be cast in the same light. The responsivity of a radiometer may be written as

$$\Re = \Re(\lambda, x, y, z, \theta, \varphi, t, T_r, T_s, T_a, E, s, p, \gamma, RH, \cdots) \ . \tag{11-4}$$

This form helps to delineate the independent variables. The dots indicate that I probably haven't thought of everything, but I tried. The measured quantity is obtained by dividing the output electrical signal by this responsivity. The responsivity is found by calibration. Now consider the variables.

The responsivity is obviously a function of the **wavelength**, λ. All detectors, even the so-called black ones, have some variation in their response with wavelength.

The responsivity generally varies with the **position**, x,y, of the incoming beam on the radiometer. In general the components have a transmission variation over their surface, as does the response of the detector.

The responsivity changes with the **angles of incidence**, θ,φ, on the radiometer. This is equivalent to a change in field angle, and every optical system has aberrations that vary with field angle. Further, the reflectivity of components is a function of incidence angle. There is a cosine projection factor as well.

The responsivity varies with **time**, t. This may be the relatively short integration time of a detector or it may be the gradual change of responsivity over time because of degradation of detectors, dirt, tarnishing of the mirror, etc.

The responsivity is a function of the **temperature of the radiometer**, T_r, since detectors change their responsivity with temperature, focal positions change, dimensions change, and background radiation changes.

The responsivity is a function of the **source temperature**, T_s. This can be because the flux level changes, the spectral distribution changes, or both.

The responsivity may change with **environmental temperature**, T_a, since the atmospheric transmission and background radiation may change.

The responsivity may change with the **incidance**, E, on the instrument. This is a function of linearity.

The responsivity may change with **polarization**, s, p, since every radiometer is polarization sensitive to some degree. See Chapter 15 for more detail on the polarization characteristics of sources and radiometers.

The responsivity may change with **degree of coherence**, γ. This may be due to a difference in the degree of coherence between the source and the calibration source.

The responsivity may change with the **relative humidity**, RH, (or the absolute humidity) of the environment.

In moments of frustration and pique it has been suggested by some of my colleagues and me that the responsivity may even change with the **phase of the moon**. (This would surely be a werewolfe effect!)

This recital of possible variations does not mean that every radiometer is affected by all these variables to a significant degree all the time. It should be viewed as a checklist to determine which of them are significant for the measurement at hand—this time.

11.4.1 THE VENUS RADIOMETER EXAMPLE

This example is described to illustrate two of these calibration dependencies, both of which are rather subtle.

Some years ago we designed a radiometer that flew to Venus and fell through its atmosphere, sampling solar radiation as it descended to obtain data on the Venerian greenhouse effect[1]. The radiometer consisted of five channels, each with a 5-degree angular diameter with sapphire windows and light pipes and silicon detectors. The probe was designed to spin slowly as it fell through the atmosphere to scan annuli in the sky and make measurements as it spiraled to its demise on the 750 K surface of the planet. The atmosphere is mostly dense carbon-dioxide vapor with local clouds of nitric and sulfuric acid.

Our design was completed and approved and the instrument was fabricated. We started calibration. The first thing we found was that in our lab the responsivity in the afternoon was different from that in the morning. The first guess was a warming effect, responsivity changing as a function of temperature. At the time we were in the Optical Sciences Annex at the University of Arizona, and the air conditioning could have been better. Further exploration, with thermocouples, indicated that this was not the effect. Could it be the after-lunch doldrums? This happens to humans but not radiometers. Ultimately we put the radiometer in a sealed, dessicator-type jar and controlled the atmosphere. Eureka! The responsivity changed with humidity. When kept dry and controlled there was no change. When wet, the responsivity dropped. When exposed to an oxygen-rich atmosphere, it increased. So, relative humidity in the variable list is not only not far-fetched, it happened. One might add to the list the general atmosphere, as in oxygen rich.

We then considered other effects. The Venerian atmosphere is a fog of carbon dioxide. Could the wings of our angular response function make a difference?

---

[1]Tomasko, M., L. Doose, J. Palmer, A. Holmes, W. Wolfe, A. DeBell, L. Brod, and R. Sholes, "Pioneer Venus Sounder Probe solar flux radiometer," IEEE Transactions on Geoscience and Remote Sensing GE-18, **93** 1980.

This is the angular effect in a different guise; it required a measurement of the angular field of view down to low levels and calculations of the input from the wings. We measured the response out to 10 degrees where it was down to 0.001 of the maximum. We integrated from 5 to 90 degrees at the appropriate levels, and found we were all right.

This same effect surely holds for the spectral passband as well.

11.4.2 SPECTRAL VARIATIONS

Another application is measuring the ultraviolet radiation from the sun. The sun is approximately a 5900 K blackbody. The peak is at about 0.5 μm and it falls off rapidly toward the ultraviolet at 0.3 μm. It is most important to have very steep filters to reject the visible light from the sun while measuring at about 0.3 μm. Whereas the Venus radiometer needed relatively sharp angular filters (well-defined field stops), ultraviolet measurements of the sun, or other sources of approximately this spectral distribution, require very sharp spectral filters. The flux from the sun between 300 nm and 3000 nm is 15,000 times that between 200 and 300 nm!

11.5 A TAXONOMY OF MEASUREMENTS

Organization can be helpful. This is my classification of the different types of radiometric measurements. It has been done differently by others[2].

The three broad classifications, the genera, seem to be measurements of flux, the measurement of the radiometric properties of materials, and the measurement of the response of instruments.

Flux measurements divide nicely into the measurements of power, incidance, exitance, intensity, and radiance. Each of these can be measured spectrally, totally, or in a band. Each can be measured angularly, hemispherically or over some large angular subtense. They can be measured as an average over time and as a fluctuation. They can be for high values of flux or low ones. They can be in a significant background or not.

---

[2]G. Zissis, "Radiometers," Chapter 20 in W. Wolfe and G. Zissis, Eds., *The Infrared Handbook*, US Government Printing Office, 1978 (available from SPIE).

Taxonomy

Flux Measurements (Spectral, Total, Bandwidth, Normalized)
    Power
    Exitance
    Incidance
    Intensity
    Radiance

Material Measurements (Spectral, Total, Bandwidth)
    Emissivity
        High Value
            High Temperature
                High Background
                Low Background
            Low Temperature
                High Background
                Low Background
        Low Value
            High Temperature
                High Background
                Low Background
            Low Temperature
                High Background
                Low Background
    Absorptivity
        High
        Low
    Reflectivity
        Specular
        Directional-Hemispherical
        Hemispherical-Directional
        Bidirectional
    Transmissivity
        Specular
        Directional-Hemispherical
        Hemispherical-Directional
        Bidirectional
    Refractive Index
        Highly Accurate
        Approximate
    Refractive Index Change with Temperature

Instruments
        Radiometers
        Fibers
        Cameras
        Sensors
        Imagers
Temperatures
        Radiation
        Ratio
        Distribution
        Color
        Differences

The properties of radiometric materials consist of transmittance, reflectance, absorptance, and emissivity. They can be measured under the same set of conditions mentioned in the taxonomy. These material properties are measured differently under different conditions. That is why there are some entries in the tree that may look illogical. Emissivity, for instance, is measured one way if the both the value and the temperature are high, so that the background is insignificant. When the reverse is true, usually the reflectivity is measured and their complementary relationship is used. Care must be taken, however, since the complementarity only applies under certain conditions (see Section 4.7).

I I .6 RECAP

This chapter has described an approach to measurement, the measurement equation, and rules for measurement. The two rules are simple: calibrate like you measure and think of every possible influence. The measurement equation helps in this by writing the responsivity as a function of all conceivable variables. A taxonomy of measurement is given. It is how I approach different measurement situations. I go down the taxonomy tree. Accuracy and precision have been described as the degree to which truth is obtained and the variability in the set of measurements generated to get that truth. It is noted that most people quote inaccuracy and imprecision, since they quote very small numbers.

CHAPTER 12

MEASUREMENT OF FLUXES

The measurement of field quantities, that is, flux, flux density, radiance, and intensity, involves the use of a radiometer, an instrument that measures the amount of flux that gets to the detector. Such a measurement requires an appropriate instrument and its calibration. In this chapter we assume that the radiometer can measure the flux with sufficient accuracy.

12.1 MEASUREMENT OF POWER

This is the basic measurement. The power on the detector creates a certain electrical signal, and *via* calibration this represents the power received by the radiometer. Power, normalized to the detector response and spectral band, can be measured by the ESRs described above, by the self-calibrated detectors described above, and by both chopper radiometers and dc radiometers described later.

12.2 MEASUREMENT OF INCIDANCE *(IRRADIANCE)*

Incidance is a flux density. If the radiometer can measure power, it can measure flux density if the area of the entrance aperture is known to sufficient accuracy. Simple division does the job. Of course, it will be the incidance averaged over the area of the entrance aperture. The measurements must be just like the calibration, and the calibration requirements and procedures are described in this chapter.

12.3 MEASUREMENT OF EXITANCE

Exitance is a source property. It can be inferred from the measurement of power. If the power on the detector is known (by calibration and measurement), then the power from the source can be calculated with certain assumptions. The main assumptions are that the distance and size of the source are known and so is the transmission of the intervening atmosphere. If the atmosphere is not known, many investigators cite an *apparent* radiant exitance. In this case, *apparent* means "without correction for atmospheric losses." This can often be very important. One such case involved understanding the emission from an ICBM. Of course, the rocket had to be measured from a distance, and it radiates copiously in the carbon dioxide band. But the atmosphere absorbs intensely in the carbon dioxide band. The apparent radiant exitance had little relation to the real values.

12.4 MEASUREMENT OF INTENSITY

Power is measured, but it is ascribed to a "point" source. The angular distribution of the radiant intensity also may be measured. The radiometer so used must be calibrated with a point source. Since a "point" source is almost always at an appreciable distance, the concept of apparent intensity applies here as well.

12.5 MEASUREMENT OF RADIANCE

Power is measured, but it is ascribed to an extended source. The angular distribution of the radiance may also be measured. The radiometer so used must be calibrated with an extended source.

12.6 CALIBRATION TECHNIQUES FOR RADIOMETERS[1]

The different types of flux measurements just discussed are usually made with radiometers. The radiometers must be calibrated just like the measurement is to be made. Thus, there are several different methods for calibrating a radiometer. These are discussed below.

12.6.1 DISTANT, POINT-SOURCE METHOD

In this method the source is smaller than the field stop and is very far away. The calibration must be done in an environment that does not affect the accuracy. The method is shown schematically in Figure 12-1.

Examples are calibrations against stars and against the sun. The sun is a point source to 1%, if the angular subtense of a detector element is larger than 33 minutes. There exist many stars the output of which is known sufficiently well that they may be considered as standard sources. Sufficiently well, in this case, is about 15%. Some calibrations are done against standard sources in the laboratory, and collimating optics are then used, as shown in Figure 12-2. The image of the source must still be smaller than the field stop, but the size of the collimating beam, its cross section, is now an additional consideration. If the beam is smaller than the radiometer entrance pupil, then not all of the radiometer

---

[1]G. Zissis, "Radiometers," Chapter 20 in W. Wolfe and G. Zissis, Eds., *The Infrared Handbook*, US Government Printing Office, 1978 (available from SPIE).

is being calibrated. If the beam is larger, then some of the light is lost, and it must be accounted for. If the beam is exactly the size of the entrance pupil, then excruciating care must be exercised in alignment. My recommendation is to make the beam enough smaller that alignment is not a problem, but still large enough that most of the radiometer aperture is filled. An additional complication, especially with some large systems and in the infrared, is the diffraction limit of the system. In such cases, a single detector element is the field stop, and it is sized to be exactly equal to the first lobe of the diffraction pattern, or the square that includes that lobe. A smaller input beam will produce a larger Airy disk, and some of the light will miss the detector. In such a case, it is better to overfill and to profile the input beam.

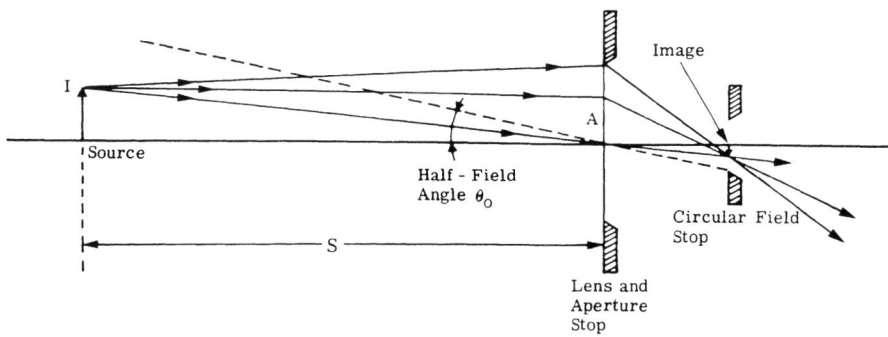

Figure 12-1. Distant small-source configuration.

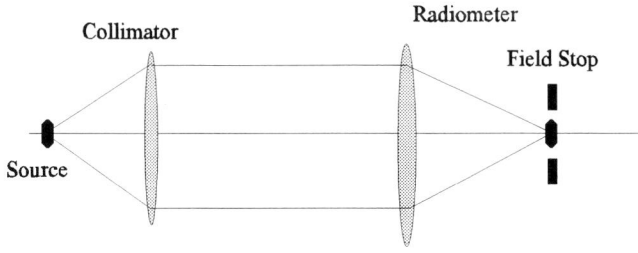

Figure 12-2. Distant small-source configuration with collimator.

12.6.2 DISTANT, EXTENDED-SOURCE METHOD

If the point source is replaced by an extended source such that its image is larger than the field of view (as shown in Figure 12-3), all the imaging properties are correct, but some of the light is lost. However, because the source is larger, this is a technique for the performance of radiance calibration. It may be compared to the distant, point-source and the near, extended-source methods. For radiance calibration it is superior. The source need be only large enough to overfill the field stop, and not the entrance pupil. The beam should be sized according to the advice for the distant point source.

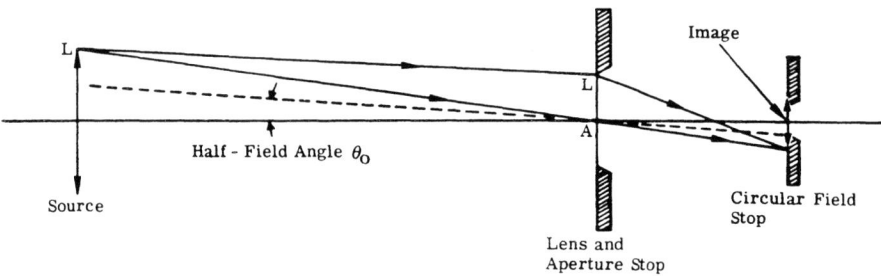

Figure 12-3. Distant extended-source configuration.

12.6.3 NEAR, EXTENDED-SOURCE METHOD

A large blackbody is placed near the aperture and fills it, in the manner shown in Figure 12-4. It clearly is a radiance calibration method that fills both stops. It does not emulate properly the rays of a distant, extended source, and it is harder to make a good, large, black, uniform source than a smaller one.

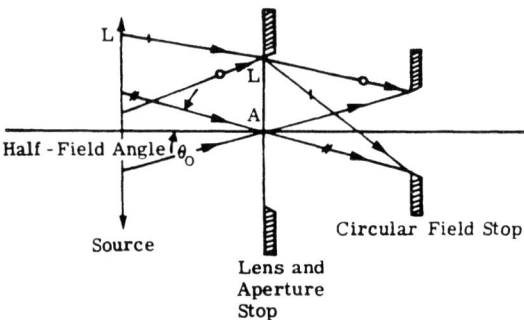

Figure 12-4. Near extended-source configuration.

12.6.4 JONES TECHNIQUE[2]

By placing a source close to the field stop, it can be fully and uniformly illuminated, thereby providing a method for radiance calibration of the radiometer. It is not necessary that the actual source be there; a good image of it will do. This means that, rather than collimating a source output, the source is imaged in that region of the radiometer. Figure 12-5 shows the requirements for the placement. The top ray shows the angular subtense of the field stop at the entrance pupil. This is the field of view of the radiometer. The bottom ray can be constructed by using the same field of view and backing the vertex away from the entrance pupil until the entrance pupil subtends the full field angle. The source is placed between these two rays as shown. It will then uniformly illuminate the field stop.

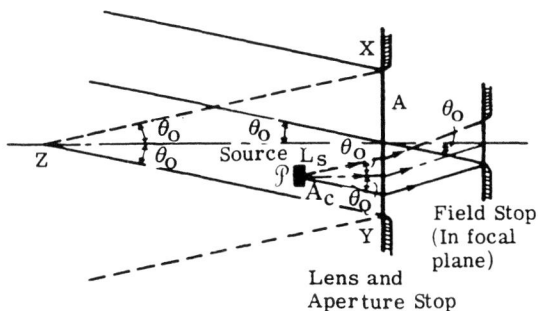

Figure 12-5. Near small-source configuration, the Jones method.

12.7 RECAP

This chapter deals with the measurement of flux in its different geometric forms. Of course, power on a detector is always measured, but the conditions of the experiment and calibration dictate whether a power, radiance, intensity, or incidance is measured. Techniques have been described for measuring point sources and extended sources.

---

[2]Wyatt, C. L., *Radiometric Calibration Theory and Methods*, Academic Press, 1978.

CHAPTER 13

MEASUREMENT OF MATERIAL PROPERTIES

These include, as described in the taxonomy, emissivities, absorptivities, reflectivities, transmissivities, and refractive indices. They are discussed in turn.

13.1 TOTAL HEMISPHERICAL EMISSIVITY[1]

This measurement is usually made by invoking some form of energy conservation. The process is to suspend the unknown in a hohlraum (a complete enclosure that is in thermodynamic equilibrium) and generate heat electrically in the unknown. The temperature difference between the unknown and the walls of the hohlraum provide the emissivity information. The equation for total radiative transfer, the Stefan-Boltzmann equation, includes the constant, σ, the emissivity, and the temperature. Thus, proper measurement of the temperatures, the flux, and auxiliary parameters can give the emissivity. Drummeter and Goldstein have described their approach in the measurement of coatings for the Vanguard satellite. Their test body, a 10-cm diameter aluminum sphere with the unknown coating, was suspended in an evacuated, 40-cm, spherical, aluminum chamber that was kept at -75 °C (198 K) with a dry-ice and alcohol bath. The arrangement is shown schematically in Figure 13-1. The inside of the chamber was coated with a Glyptal black, and electrical power was supplied to a cartridge heater in the sample. Both temperatures were measured with thermocouples. Both the power leads and the thermocouple leads to the unknown were maintained as thin as possible to minimize heat loss. The inside of the chamber was maintained at 10^{-5} torr. The power transfer is given by

$$\Delta\Phi = EI = \frac{\varepsilon_1\varepsilon_2\sigma A_1 A_2(T_1^4 - T_2^4)}{\varepsilon_2 A_2 + \varepsilon_1 A_1 - \varepsilon_1\varepsilon_2 A_1} + K(T_2 - T_1) + \frac{CdT}{dt} , \qquad (13\text{-}1)$$

where Φ = input power [W], E = heater voltage [V], I = heater current , ε_1= hemispheric surface emissivity, ε_2= emissivity of the walls, σ = Stefan-Boltzmann constant, A_1 = sample surface area, A_2 = chamber surface area, T_1 = sample temperature [K], T_2 = chamber temperature [K], K = conductive loss constant, and C = heat capacitance of the sphere.

---

[1]Drummeter, L. F. and E. Goldstein, "Vanguard emittance studies at NRL," in F. J. Claus, ed., *Surface Effects on Spacecraft Materials,* Wiley, p. 152, 1960.

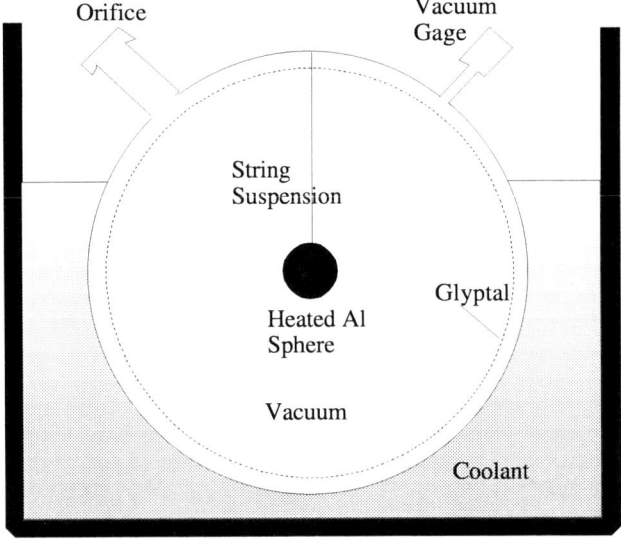

Figure 13-1. The total hemispherical chamber of Drummeter and Goldstein.

The first term represents radiative heat interchange (refer to the Appendix for this interaction factor), the second is the heat that is conducted between the chamber and the sphere through the wires and the support, and the last is the heat storage in the sample sphere. Under conditions of equilibrium, the last term vanishes, and

$$\Delta\Phi = EI = \frac{\varepsilon_1 \varepsilon_2 \sigma A_1 A_2 (T_1^4 - T_2^4)}{\varepsilon_2 A_2 + \varepsilon_1 A_1 - \varepsilon_1 \varepsilon_2 A_1} + K(T_2 - T_1) \; . \qquad (13\text{-}2)$$

On the assumption that A_1/A_2 is small enough, the relationship is

$$\Delta\Phi = EI = \varepsilon_1 \sigma A_1 (T_1^4 - T_2^4) + K(T_2 - T_1) \qquad (13\text{-}3)$$

and

$$\varepsilon_1 = \frac{EI - K(T_1 - T_2)}{\sigma A_1 (T_1^4 - T_2^4)} \ . \tag{13-4}$$

Thus, the emissivity can be found from a knowledge of the input electrical power, the thermal conductance, the temperatures, and the area of the spherical sample, as claimed. The chamber is a 40-cm sphere and the test article is a 10-cm sphere, so the area ratio is 6.25%. If the Glyptal emissivity is close enough to unity, then the error is even smaller. For an emissivity larger than 0.95, the error is smaller than 0.355%.

In some cases the conduction is negligible. At any rate, the emissivity can be evaluated in terms of the input electrical power, the Stefan-Boltzmann constant, the area of the sphere, and the two temperatures. The heating has been by way of electrical resistance. A different method must be used for insulators.

13.2 SPECTRAL DIRECTIONAL EMISSIVITY

Richmond[2] has reported on instrumentation used by the NBS (NIST) for the measurement of total, normal emissivity of secondary standard strip lamps. The basic technique is to compare the emission of the lamps to that of a blackbody at the same temperature. The receiver is a monochromator with one of several different detectors: photomultiplier, lead sulfide, or thermocouple. For comparisons in the 0.5 to 2.6 µm region quartz prisms are used, but a sodium chloride prism is used for longer wavelengths. He notes that the sample can actually be anything that is heatable by an electric current. The blackbody cavity has a ratio of aperture area to internal surface area of 0.003, and the assertion is made that this is sufficient to ensure an effective emissivity of at least 0.999. The industrious, or skeptical reader, will check this against the Gouffé treatment in Chapter 6. The maximum temperature of the cavity is 1400 K. There is no direct description of how the temperature of the specimen is monitored, although the lamp current is "adjusted to a predetermined value." Presumably the other samples would be measured with a thermometer appropriate to the temperature. In making

---

[2]Richmond, J. C., "Some methods used at the National Bureau of Standards for measuring thermal emittance at high temperatures," in H. Hammond and H. Mason, eds., *Precision measurement and calibration: radiometry and photometry*, NBS special Publication 300, 7, 182 1971.

measurements of this type, care should be exercised to ensure that the reflected light does not influence the measurement. A short analysis will illustrate this.

Assume that the sample has isotropic reflectivity and emissivity, that the room is at 300 K and has an emissivity of one. Then the total radiance from the sample is the sum of the emitted and reflected components:

$$L_{total} = \varepsilon L^{BB}(T_{sample}) + (1-\rho)L^{BB}(T_{room}) . \tag{13-5}$$

A more exact representation includes the sum of the different parts of the room in the following way:

$$L_{total} = \varepsilon L^{BB}(T_{sample}) + \frac{1-\rho}{\pi}\Sigma_i \varepsilon_i L^{BB}(T_i)\Omega_{p\cdot i} , \tag{13-6}$$

where ρ is the hemispherical reflectivity of the sample and the subscript i represents the different room emitters, each with its own temperature T_i, emissivity ε_i, and projected solid angle Ω_{pi}. The first approximation will be used to show the approximate magnitude of the various contributors and errors. For the entire spectral band from 0.5 to 2.6 μm, as indicated by Richmond, the values are

$$L_{total} = \varepsilon[2.85 - 1.63\times10^{-7}] + 1.63\times10^{-7} . \tag{13-7}$$

The background contribution is surely negligible in this case. It must also be negligible for any quasi-monochromatic component. However, as the specimen temperature is reduced, the background contribution becomes more important. Note also that as the emissivity is lowered the background contribution also becomes more important. For this case, however, even a very diffuse sample with 98% hemispherical emissivity would still be measured in error by only 2.85×10^{-6}. But, for a sample temperature of 1000 K, the error is 1.63×10^{-7} parts in 0.33ε ($4.9\times10^{-7}/\varepsilon$,) and for 500 K the value is $3.3\times10^{-4}/\varepsilon$. For a spectral band that extends further into the infrared, the error also increases, as more of the 300 K background radiation is accepted.

McDonough[3] has described instrumentation for the measurement of emissivity of materials at a temperature of 360 K in the spectral region from 4.0 to 13.5 μm. The basic feature is to chop the radiation from both the heated source and the background and use subtractive correction techniques. The chopped radiation passes through a monochromator to a thermocouple, its amplifier, and to an oscilloscope. Although he comments on the problems with the small differences between the background and the signal, he does not provide a detailed analysis. That is an exercise left for the student! There usually is a problem in this technique that he does not describe. The front surface of the chopper reflects the background, and the sample still both emits and reflects. Great care in calculating the radiative transfer is in order.

13.3 TOTAL DIRECTIONAL EMISSIVITY

The conversion to total emissivity from a spectral one is in the detector. Replace the monochromator by a thermal detector. Watch even more closely for background contributions. It is true that no detector is completely black over all wavelengths, but this approach should be reasonably accurate.

13.4 SPECULAR REFLECTIVITY

The three main techniques for the measurement of specular reflectivity are those of Strong[4], Bennett and Koehler[5], and substitution, which really does not have any name associated with it.

13.4.1 SUBSTITUTIONAL METHOD

This can be one of the most accurate and simple methods of reflectance measurement. Shine a beam of light onto a detector. The beam should be modulated for several reasons: to separate the measurement beam from the background, to move the detection away from dc and resultant excess noise, and

---

[3]McDonough, R., in F. J. Claus, Ed., *Surface Effects on Spacecraft Materials,* p. 141, Wiley, 1960.

[4]Strong, J., *Procedures in Experimental Physics*, Prentice Hall, 1938.

[5]Bennett, H. E. and W. F. Koehler, "Precision measurement of absolute specular reflectance with minimized systematic errors," Journal of the Optical Society of America **50**, 1, 1960.

to provide for narrow-band detection. Then move the detector to another position and interpose the mirror to be tested. The arrangement is shown schematically in Figure 13-2.

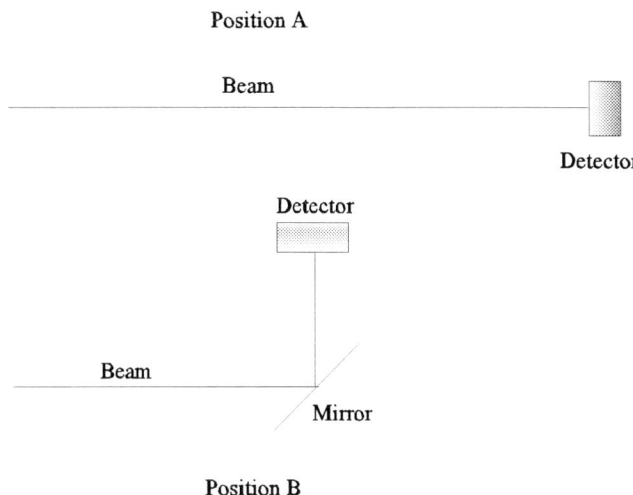

Figure 13-2. The substitution method for measuring specular reflectance.

The ratio of the measurements is the specular reflectivity of the mirror. The biggest caveat is that everything must remain constant during the two measurements. This means that the beam must not change; the same area of the detector must be illuminated (perhaps the whole thing); the geometry must all be the same. In the simplest realization, the detector is moved, and maximum signal is sought in both the before and after situations. One realization of this was carried out in our lab, using an electro-optical modulator with feedback [6]. By constantly monitoring the output of this laser with a detector, we found that the modulator system controlled the beam to be constant at some value for ten minutes or more and with a variation of less than 0.02%. Such a system could then be used with the substitution as long as the movements could be done within ten minutes. They could. A search was always done for the maximum signal, and, since the beam was collimated with very little divergence, the position along the optical axis was not critical. The experimental setup is shown in Figure 13-3. Other variations

[6]Lee, S. M., *Investigation and Extension of Self-Calibration Radiometry*, Dissertation, The University of Arizona, 1983.

on this theme include the possibility of calibrating two detector responsivities *with respect to each other* with this sort of accuracy at this power level. Then the experiment can be performed using the two detectors, one with the mirror and the other without. This goes much quicker. Zalewski and Geist have also discussed some other geometries[7].

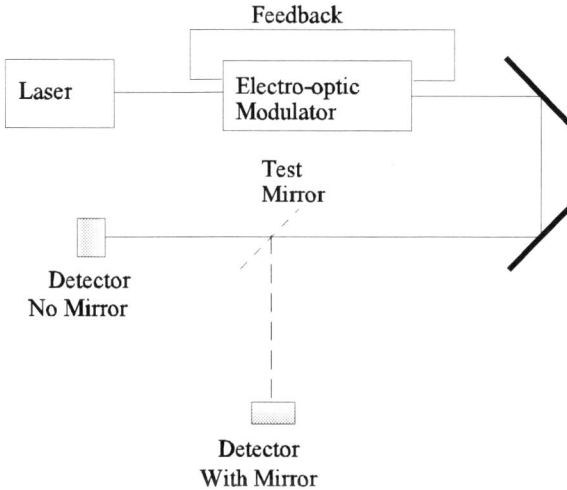

Figure 13-3. The University of Arizona setup for measuring reflectivity to calibrate a self-calibrating detector.

13.4.2 THE STRONG METHOD

This technique, also called the VW method, is a substitutional one as well. A device is arranged, as shown in Figure 13-4, with an appropriate source beam entering from the lower left. This beam may be monochromatic or cover a wider spectral range. Part of the apparatus is a good specular mirror, labeled the instrument mirror in the figure. The beam is reflected off this mirror when it is in

---

[7]Zalewski, E. and J. Geist, "Solar cell spectral response characterization," Applied Optics **18**, 3942, 1979; "Status of detector response transfer capabilities regarding accurate photometry," Journal of the Optical Society of America **68**, 1389, 1978.

the upper position, A, and then to an appropriate collection and detection apparatus toward the lower right. Then the sample cell is rotated and the sample is inserted so that the light now reflects off the same instrument mirror and twice from the sample. The reflectance is then the square root of the ratio of the values. If the source is unchanged during the two measurements, the detector first reads $\rho_i \Phi$ where ρ_i is the instrument mirror reflectivity and Φ is the flux from the source. Then the detector reads $\rho_i \rho^2 \Phi$. The ratio is just ρ^2, as claimed.

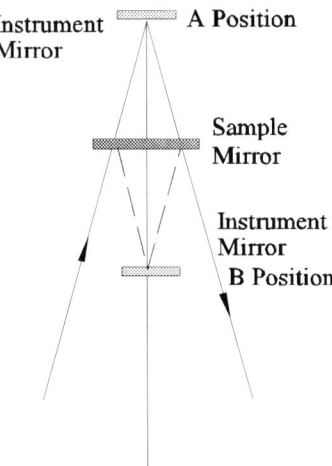

Figure 13-4. The Strong VW technique. The mirror is used in the A position to measure the instrument mirror reflectivity. The holder is rotated to the B position and the sample inserted.

13.4.3 The Bennett-Koehler Method

This is a much more elaborate equipment. It can test samples at a somewhat larger range of angles of incidence. It can also, in a sense, calibrate itself and is relatively insensitive to tips of the detectors. Consider the paths of the light, shown in Figure 13-5, because that is the secret of the good performance. The light enters from the upper left, perhaps from a good spectrometer. The first mirror M_1 serves with mirror M_7 to demagnify the output slit of the monochromator and image it on the sample. The intermediate mirrors are all flats, and serve mostly the master of convenience. After the beam is reflected from the sample, it goes to mirror M_8 and is refocused back on the sample by that mirror, from whence it goes to M_9 and there to the output optics: $M_{13} \dots M_{16}$ and then $M_{18} \dots M_{21}$. Note that there have been two reflections from the sample, providing the same extra sensitivity as in the Strong technique. When the sample is removed, the beam from M_7 goes to M_{11} and then back to M_9, after which it follows the same path. If it can be shown that M_{11} is identical to M_8, then an excellent measurement has been made. The paths are otherwise identical, including length and number of surfaces. But, for the most accurate measurements, it may *not* be assumed that M_{11} is identical to M_8. So an auxiliary measurement is made. The sample chamber, which consists of the sample and mirrors M_6 and M_{13}, is rotated 180 degrees; the light follows the path shown with dotted lines. Sample in and sample out are done again. This time mirrors M_{12} and M_{10} replace M_7 and M_9 and M_{11} and M_8 reverse roles.

Shaw and Blevin have described a technique for measuring both normal, specular reflectance and normal, specular transmittance[8]. It may involve the switching of two transparent plates of the system.

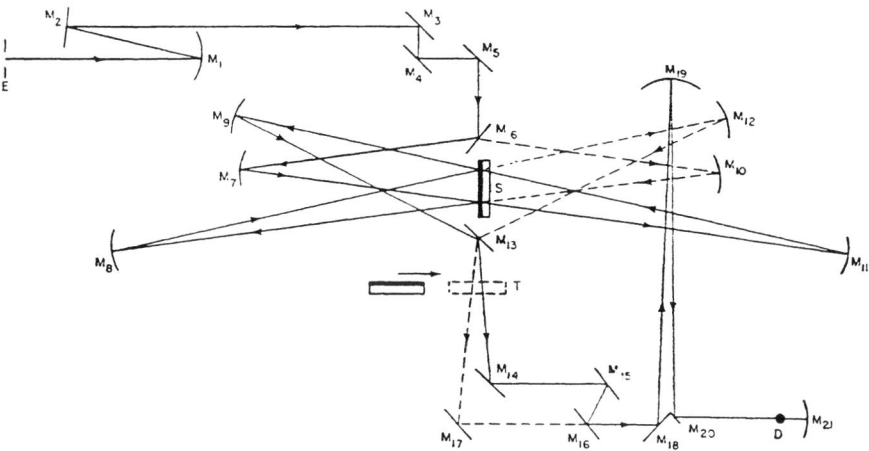

Figure 13-5. The Bennett-Koehler reflectometer.

I 3.5 SPECULAR TRANSMITTANCE

The quantity is defined as the ratio of the power in a beam after it has been transmitted through a sample to the same value in the incident beam. It includes the effects of the surface reflections as well as the bulk absorption. It may be the most straightforward of the measurements in this section. The substitutional

---

[8]Shaw, J. E. and W. R. Blevin, "Instrument for the absolute measurement of direct spectral reflectances at normal incidence," Journal of the Optical Society of America **54**, 334, 1964.

reflectivity technique can be used, but is even simpler. Shine a beam of light on a detector; insert the plane parallel plate, and calculate the ratio of the signals. There is at least one thing to worry about: The insertion of the plate will affect the position of focus of a converging beam. The difference in focus for a ray that enters the plate at an incidence angle of θ is

$$\Delta f = d\cos\theta \left[\tan\theta - \frac{n\sin\theta}{\sqrt{1 - n^2\sin^2\theta}} \right] , \qquad (13\text{-}8)$$

where d is the thickness of the plate and n is the refractive index of the plate relative to the medium outside. This ratio of the power in the transmitted beam to that of the incident beam is called the *external transmittance*, τ_{oo}. It is calculated to be

$$\tau_\infty = \frac{(1-\rho)^2\tau_i}{1 - \rho^2\tau_i^2} , \qquad (13\text{-}9)$$

where τ_i is the internal transmittance. The infinity subscript is a reminder that this transmittance takes into account an infinite number of reflections from surface to surface and out. The internal transmittance is given by Bouguer's law,

$$\tau_i = e^{-\alpha x} , \qquad (13\text{-}10)$$

where α is the absorption coefficient and x is the thickness of the plate.

The reflectivity that one calculates for this same plane parallel plate and with an infinite number of reflections is

$$\rho_\infty = \rho + \frac{(1-\rho)^2\tau_i^2}{1 - \rho^2\tau_i^2} . \qquad (13\text{-}11)$$

By using the fact that

$$\tau_\infty + \rho_\infty + \alpha_\infty = 1 \qquad (13\text{-}12)$$

one can find (with quite a bit of algebra) that

$$\alpha_\infty = \frac{(1-\rho)(1-\tau_i)}{1-\rho\tau_i} \; . \tag{13-13}$$

Many investigators have approximated Eq. 13-9 by using just the numerator, i.e.,

$$\tau_x = (1-\rho)^2\tau_i = (1-\rho)^2 e^{-\alpha x}. \tag{13-14}$$

The relative error in using this approximation is

$$RE = \rho^2\tau^2 \; . \tag{13-15}$$

For a piece of glass with 5% reflectivity and 95% internal transmittance, this error is about 0.2%, but for an equally nonabsorbent piece of germanium that has a reflectivity of 34%, the error is about 10%.

The internal transmittance, which leads to an absorption coefficient, cannot be measured directly, but must be inferred from the measurement of external transmittance. The usual process has been to use the approximate equation, Eq. 13-14, and invert it:

$$\tau_i = \frac{\tau_x}{(1-\rho)^2} \; . \tag{13-16}$$

The relative error can be found either by taking the derivative of the log of the expression or by taking the derivative directly and dividing by the expression. The result is

$$RE = \frac{d\tau_x}{\tau_x} - \frac{2d\rho}{1-\rho} \; . \tag{13-17}$$

The measurements can usually be made to about 1% so that the relative error is the rms sum of the error in measuring reflectivity and in measuring the external transmittance, and is

$$RE \approx \sqrt{0.01^2 + \left(\frac{0.02}{0.9}\right)^2} = 0.0244 \qquad (13\text{-}18)$$

for a reflectivity of 90%. It is reasonable to conclude that specular transmittance measurements can be made at the level of about 1%. The uncertainty can be lowered with very careful work.

13.6 INTERNAL TRANSMITTANCE AND ABSORPTION COEFFICIENT

The internal transmittance is determined to learn the absorption properties of a sample. This property is usually described by the absorption coefficient, α. We can learn something about how to measure, i.e., calculate, the absorption coefficient from the internal transmittance by an error analysis of the Bouguer law. The relation between internal transmittance and absorption coefficient is

$$\tau_i = e^{-\alpha x} . \qquad (13\text{-}19)$$

The relative error is found by the derivative of the logarithm and yields

$$\frac{d\tau_i}{\tau_i} = \alpha dx + x d\alpha . \qquad (13\text{-}20)$$

I have ignored the negative sign and gone directly to the rms value,

$$d\alpha = \sqrt{\left(\frac{d\tau_i}{x\tau_i}\right)^2 + \left(\alpha\frac{dx}{x}\right)^2} . \qquad (13\text{-}21)$$

The uncertainty in determining α can be viewed as a straight line that has an intercept given by the first term and a slope given by the relative uncertainty in measuring the thickness. The uncertainty in inferring internal transmittance is estimated to be about 1%.

The first term, which can be viewed as the x-axis intercept, can be made smaller by making the relative error in the transmittance measurement smaller, and by increasing the thickness. I assumed a 1% error and a 1-cm thickness. It will be difficult to decrease the error in the measurement by much, but the sample can be thicker, say 10 cm. Then the intercept will drop by an order of magnitude. The relative error, the slope, will also change, and probably be decreased. I had assumed two different relative errors of thickness measurement, 0.001 and 0.0001. This means that the error in measuring the 1-cm sample is 0.001 and 0.0001 cm (10 μm and 1 μm). They would probably have to be done interferometrically, and the surfaces would have to be parallel and flat to a few waves. If the 10-cm sample is used, one could make the same interferometric measurements and get 10 times the accuracy. This treatment predicts obtaining the absorption coefficient with an uncertainty from 0.1% to 1%, but I did it with the approximate equation. Unfortunately, the analysis does not work this easily when the exact equation is used, and there are materials for which the approximation is too approximate.

The ratio of transmittances for samples of two different thicknesses turns out to be

$$\frac{\tau_1}{\tau_2} = e^{-(\alpha x_1 - \alpha x_2)} \left| \frac{1 - \rho^2 e^{-\alpha x_1}}{1 - \rho^2 e^{-\alpha x_2}} \right|. \tag{13-22}$$

One approach based on this technique bears discussion. It might be called that of differential spectroscopy. Put a thin sample in the reference beam of a dual-beam spectrometer, and a thicker sample in the measurement beam. The spectrometer measures the ratio of the two samples.

The approximate equation. Eq. 13-14, can be used for many cases, but not always, and it is important to evaluate when and where it is applicable.

It is important to eliminate the reflectivity, not just to calculate it. One could argue that the refractive index is known to about five significant figures, so the reflectivity can be calculated with great accuracy. That is true, but the surface reflectivity depends on the surface film, the dirt, the oxidation, the fingerprints, etc. Thus, the experimental elimination technique is of the essence.

The most widely used method of measuring the absorption coefficient of very transparent materials is what might be called adiabatic, laser calorimetry. A laser is used as a power source. The sample absorbs some of the incident power, and increases its temperature (according to its absorptivity, specific heat, and mass). The temperature increase is measured with one or more thermocouples usually placed on the periphery of the sample. The measurement can also be made based on the rate of temperature rise. In both senses this is much akin to absolute radiometry; heating and temperature changes are measured and related to other physical quantities. In absolute radiometry, great care is exercised in absorbing all the incident radiation (with cones, disks, blacks, etc.), but in this case the proportion of energy absorbed is important. It is related to the internal transmittance and therefore the absorption coefficient. The operational equation is given as

$$\alpha = \frac{mc_p}{x\Phi} \frac{2n}{n^2+1} \left[\left(\frac{dT}{dt}\right)_{gain} + \left(\frac{dT}{dt}\right)_{loss} \right],$$
(13-23)

which is only valid when $\alpha x \ll 1$. This equation requires not only that the absorption coefficient is very small, but that the entire sample experiences the same rate of temperature change. Generally, each sample will experience a 1-$\exp(t/t_1)$ rise in temperature and a corresponding $\exp(-t/t_2)$ decay, where t_1 and t_2 are the rise and decay times, respectively.

This result may be approached several ways. The temperature increase due to specific heat is given by

$$mc_p \frac{dT}{dt} = \Phi_a = \alpha \Phi_i,$$
(13-24)

where dT/dt is the time rate of temperature rise, m is the mass of the sample, c_p is the specific heat, and α is the absorptance (not the absorption coefficient). It has already been shown that, for a plane parallel plate,

$$\alpha_\infty = \frac{(1-\rho)(1-\tau)}{1-\rho\tau},$$
(13-25)

where ρ is the single-surface reflectivity and τ is the single-pass transmittance. When the transmittance is very close to 1, then $\rho\tau = \rho$ and $e^{-\alpha x} = 1 - \alpha x$. Then

$$\alpha_{\infty} = \frac{(1-\rho)(1-\tau)}{1-\rho\tau} = \alpha x .$$

(13-26)

Therefore, the nice, simple result appears, and it is valid for these very transparent samples.

When a pulse of energy is incident on the sample, there will be a temperature rise above equilibrium, and when the pulse ends, there will be a decay back to equilibrium. Therefore, the full equation, because all the terms are the same, involves two time-temperature terms:

$$\left(\frac{dT}{dt}\right)_{total} = \left(\frac{dT}{dt}\right)_{rise} - \left(\frac{dT}{dt}\right)_{decay} .$$

(13-27)

Usually the time constants are such that a respectable exponentially shaped rise occurs, followed by a similar decay. In this analysis convective, conductive, and radiative effects have been ignored. For sufficiently short rise time, involving samples of low mass and with little thermal conduction to a sink, these assumptions do not degrade the accuracy. Surface absorption and scatter can, however, be problems. Scatter from the sample to the thermocouple appears as an anomalously fast rise and subsequent decay. It happens with the speed of light—30 cm in 1 ns! This is much faster than any thermal time constant. This level can be subtracted before the radiometric equation is used[9]. Surface absorption is not so easily handled. The single-pass absorption can then be determined from the difference between the measurements, it is said. The usual technique is to measure two samples of different thicknesses, assuming the surfaces are identical.

Pinnow and Rich[10] have used calorimetric techniques to measure the transmission

---

[9]Nistor, L. C., S. V. Nistor, V. Teodorescu, E. Cojocaru, and I. Mihailescu, "Calorimetric absorption coefficient measurements using pulsed CO_2 lasers," Applied Optics **18**, 3517, 1979.

[10]Pinnow, D. A. and T. C. Rich, "Measurement of the absorption coefficient in fiber optical waveguides using a calorimetric technique," Applied Optics **14**, 1258, 1975.

characteristics of optical fibers and found fiber-optic coefficients as low as 2.3 dB/km (1.3×10^{-5} cm^{-1}) by inserting the fiber inside a laser cavity.

The technique I like the best is that of Stierwalt and Potter[11]. By measuring the effective, multiple-pass emissivity and the sample thickness, the absorption coefficient is readily obtained. Their technique is to carefully measure the radiance from the sample and compare it with the radiance from a blackbody at the same temperature. The instrumentation and the surround must be at a temperature that is sufficiently low compared to that of the sample so that emission from the sample and not reflection of the surround is measured. This technique has the promise of reaching noise equivalent absorption coefficient levels of 10^{-9} cm^{-1} but light scattering has so far kept the measurements in the 10^{-5} cm^{-1} regime.

13.7 DIRECTIONAL-HEMISPHERICAL REFLECTION

By the Helmholtz reciprocity principle, the hemispherical-directional reflectivity is the same as the directional-hemispherical reflectivity, and it is the latter that is usually measured. There are several techniques that have been used. These include the Coblentz hemisphere, the paraboloidal reflector, the integrating sphere, and the Gier-Dunkle cavity.

13.7.1 THE COBLENTZ HEMISPHERE

The hemisphere is shown in Figure 13-6. Light enters through a slit and is incident upon the sample, from which it scatters in all directions, although not necessarily uniformly. The overlying hemisphere is reflective and returns the light to the detector. Since both the sample and the detector are placed close to the center of curvature, all of the light from the sample is reflected back to the detector. The hemisphere must be calibrated for its reflectivity in some way. The use of a standard will do it, and so will an independent measurement of the specular reflectivity of the surface. There is a certain amount of aberration caused by the fact that the detector and sample cannot both be placed at the center of curvature. A more serious problem, however, is the cosine effect on the detector. Light at large angles is not sensed with as high a responsivity as light from small angles. Some designs for cosine detectors have been tried.

---

[11]Stierwalt, D. L. and R. F. Potter, *Proceedings of the International Conference on the Physics of Semiconductors*, The Institute of Physics and the Physical Society, London, 513, 1962.

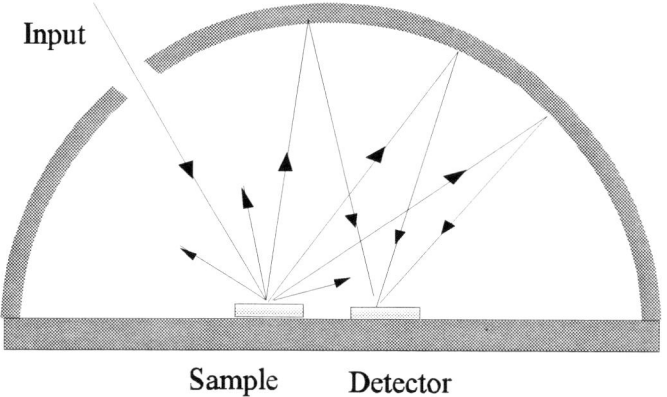

Figure 13-6. The Coblentz hemisphere.

The incidence angle can be adjusted by moving the source up and down in the slit. The spectrum is adjusted by the choice of source and detector. The source can be any of many lasers, a spectrometer, a blackbody, tungsten lamp, etc. Thus, spectral, average, or total directional hemispherical reflectivity can be measured with this instrument.

13.7.2 THE PARABOLOIDAL REFLECTORS

Figure 13-7 shows this device. Its spectral extent is the same as that of the Coblentz sphere. Although the diagram from the reference shows a collimated, monochromatic beam, the monochromaticity is not required. The incidence angle is adjusted by the position of the plane mirror. Although in this system the sample is imaged to the detector by the paraboloids, and the aberrations are thereby reduced, there is still the angle factor with which to deal.

13.7.3 THE INTEGRATING SPHERE[12]

A full sphere can be used. As shown in Figure 13-8, the detector is placed on the

[12] D. Goebel, "Generalized integrating sphere theory," Applied Optics **6**, 125, 1967; J. Jacquez and H. Kuppenheim, "Theory of the integrating sphere," Journal of the Optical Society of America **45**, 460, 1955; B. Hisdal, "Reflectance of perfect diffuse and specular samples in the integrating sphere," ibid.,**55**, 1122, 1965; D. Goebel, B. Caldwell, and H. Hammond, "Use of an auxiliary sphere with a spectroreflectometer to obtain absolute reflectance," ibid., **56**, 783, 1966.

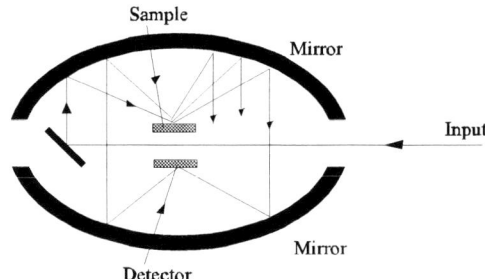

Figure 13-7. Paraboloidal reflectometer.

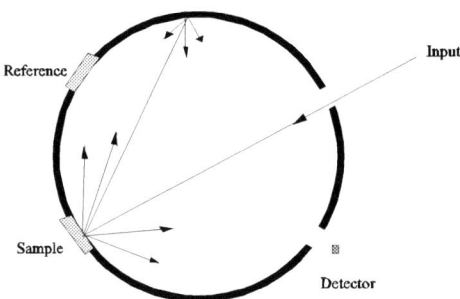

Figure 13-8. The integrating sphere.

surface of the sphere, preferably at an angle of ninety degrees or more from the sample. There are then several options for making the measurements. The sample and the reference can be placed in the same position in time sequence. They can be placed as shown in the diagram from the reference.

13.7.4 THE GIER-DUNKLE CAVITY[13]

Shown in Figure 13-9 is a cavity that is neither hemispherical, spherical, or doubly paraboloidal that can also be used. The Gier-Dunkle technique is to view a portion of the interior and then the sample.

---

[13]Gier, J., R. Dunkle, and J. Bevans, "Measurement of absolute spectral reflectivity from 1.0 to 15 microns," Journal of the Optical Society of America **44,** 558, 1954.

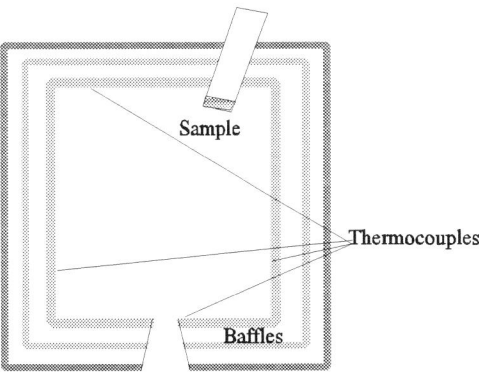

Figure 13-9. The Gier-Dunkle hohlraum.

13.8 DIRECTIONAL-HEMISPHERICAL TRANSMITTANCE

I have not been able to find any references for anyone making these measurements, and I cannot think of any important applications that require them. However, consider possible implementations. Any of the directional-hemispherical methods could be used if the transparent sample is backed by a known, specular mirror of high reflectance.

13.9 BIDIRECTIONAL REFLECTANCE AND TRANSMITTANCE[14]

This most fundamental of the methods of measuring transmittance and reflectance uses as much geometry as one can imagine. Figure 13-10 is a diagram of the instrument we built at The University of Arizona in 1979[15]. There are a number of features that are worth discussing. The light beam, usually a laser for reasons that will be discussed, illuminates a sample that is mounted on a gimbal, as shown

---

[14]Brooks, L. and W. Wolfe, "Microprocessor-based instrumentation for bidirectional reflectance distribution function (BRDF) measurements from visible to far infrared (FIR)," Proceedings of SPIE **257**, 177, 1980; F. Bartell, E. Dereniak and W. Wolfe, "Theory and of BRDF and BTDF," ibid. 154.

[15]Brooks, L., *Microprocessor-Based Instrumentation for BSDF Measurements from Visible to FIR,* Dissertation, The University of Arizona, 1982.

in Figure 13-11. For a single position of the sample, the detector can swing a full 360 degrees, allowing the measurement of both bidirectional reflectance and transmittance. Then, as the sample is moved about a vertical axis (the plane of incidence is horizontal), the same 360-degree scan can be made for a variety of angles of incidence—in the plane of incidence. Then, the sample can be tilted about a horizontal axis to change the out-of-plane angle. In this way the entire overlying hemisphere can be covered for either a transparent sample or a reflecting sample. This is the geometry that most investigators have used, although it does require certain angle transformations (Euler-angle equations) in some cases.

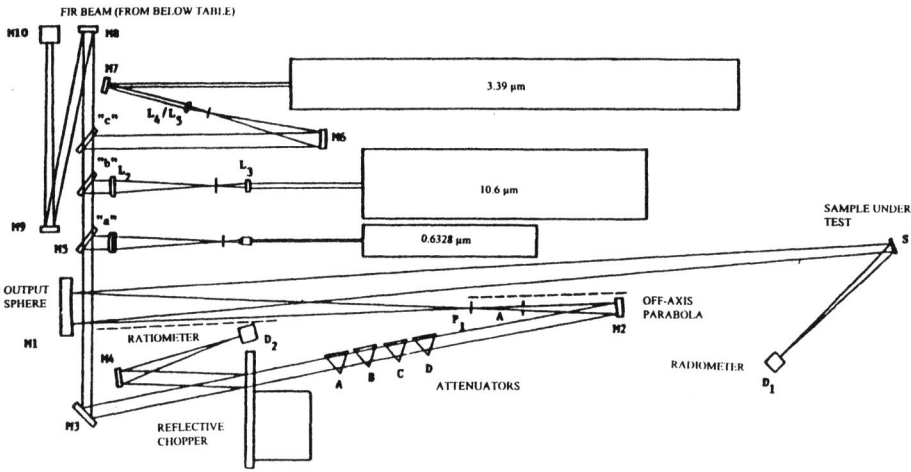

Figure 13-10. Layout of the AZSCAT bidirectional reflectance and transmittance instrument.

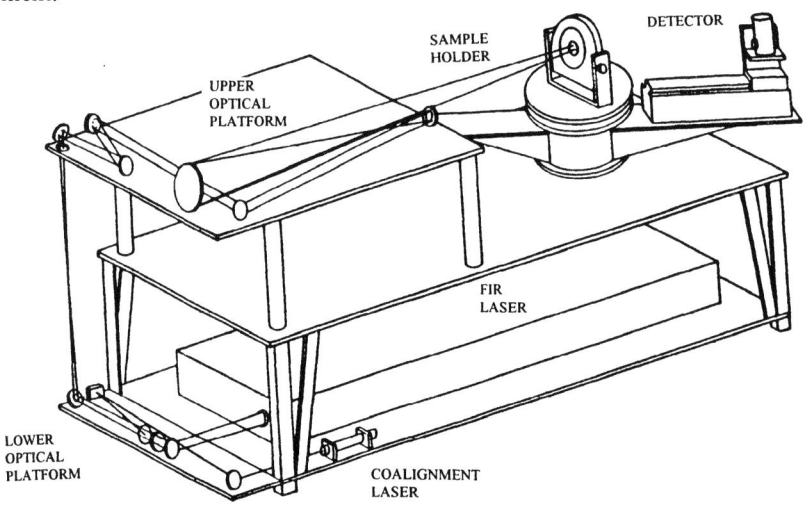

Figure 13-11. Sample-holder gimbal mounted on optical table.

It is both instructive and illuminating to calculate the sensitivity of such a system. This is done in terms of the noise equivalent bidirectional reflectivity (NEBDR). It starts with the expression for the signal-to-noise ratio in terms of the detector specific detectivity, $D*$:

$$SNR = \frac{D^* \Phi_d}{\sqrt{A_d B}} , \tag{13-28}$$

where Φ_d is the power on the detector, B is the effective noise bandwidth, and A_d is the area of the detector. The power on the detector is given by

$$\Phi_d = L \frac{A_d \cos\theta_d A_s \cos\theta_s}{R^2} = L A_s \cos\theta_s \Omega_{ds} , \tag{13-29}$$

where A_s is the illuminated area of the sample, θ_s is the angle the surface normal makes with the line of sight between the detector and the sample, and θ_d is the equivalent angle for the detector. The distance between the two is a constant R, and the solid angle the detector subtends at the sample is Ω_{ds}. The radiance L is the bidirectional reflectance times the incidence on the detector,

$$L = \rho_{bd} E_i = \frac{\rho_{bd} \tau \Phi_l}{A_i} , \tag{13-30}$$

Φ_l is the laser source power, and τ is the transmittance of the system between the source and the sample. When all of these are combined, the SNR can be written

$$SNR = \frac{D^* \rho_{bd} \tau \Phi_l \cos\theta_s \Omega_{ds}}{\sqrt{A_d B}} . \tag{13-31}$$

In the traditional way, the SNR can be set equal to 1 and the equation solved for the bidirectional reflectance, *viz.*,

$$NEBR = \frac{\sqrt{A_d B}}{D^* \Phi_l \tau \cos\theta_s \Omega_{ds}} . \tag{13-32}$$

The nominal values for our machine operating at 10.6 μm are $\Phi_I = 10$ W, $\tau = 0.5$, $B=1$, $\sqrt{A_d}=100$ μm$= 0.01$ cm, $D^*=1\times10^{11}$, and $R = 50$ cm. So, at normal incidence one has

$$NEBDR = \frac{\sqrt{A_d}B}{D^*\Phi_I\tau\cos\theta_s\Omega_{ds}} = \frac{0.01\times1}{10^{11}\times0.5\left(\frac{0.01}{50}\right)^2} = 5\times10^{-6}\,[sr^{-1}]\,.$$

$$(13\text{-}33)$$

We found experimentally that this calculation is about right. We also found that the value was not quite as good for shorter wavelengths, at 0.6328 μm, where we used a HeNe laser and a silicon detector. Note that the difference is in the $D^*\Phi_I$ product. For the 10 μm system, the product was 10^{12}. For the visible, though, the D^* is not quite two orders more and the laser is three orders less. So the result is about 20 or 30 times less sensitive. This is one of the very few applications that I have found such that the infrared is more sensitive than the visible.

The calibration of a bidirectional measurement instrument can be done either with a sample or without. The latter is accomplished by simply shining the laser light onto the detector with no intermediate sample. (This is one reason to design an instrument that covers 360 degrees.) The trouble with this procedure is that the levels of the calibration and those of measurement are very different, orders of magnitude different, and this violates rule one of calibration. The two varieties of sample calibration are a lambertian sample and an arbitrary reference that is used in a substitutional mode; first one is measured, then the other. The latter satisfies calibration rule one, but requires that the reference is known for all combinations of angles, and that is a four-fold infinity! The lambertian technique requires a lambertian sample, and there is no such thing as a lambertian reflector. However, flowers of sulfur and Halon are very good references in the visible and near-infrared and gold-plated sandpaper is good in the infrared, beyond about 5 μm[16].

13.10 REFRACTIVE INDEX

The (real part of the) refractive index is one part of the complex refractive index that consists of both the real and imaginary parts. The imaginary part, κ, is alternately called the extinction coefficient and the absorption coefficient, α. They are a little different; $\alpha=4\pi\kappa/\lambda$. Techniques have already been described for the measurement of the absorption coefficient. This section describes the

---

[16]Stuhlinger, T., *Bidirectional Reflectance Distribution Function (BRDF) of Gold-Plated Sandpaper*, Thesis, The University of Arizona, 1981.

measurement of the real part, the ratio of the speed of light in a vacuum to that in a material.

The classic technique is minimum deviation[17]. Other geometries of the prism can be used. One of these is normal incidence[18]. The minimum deviation technique has been used by NBS (NIST) over the years, and they were able to make measurements in the visible and infrared with an uncertainty of a few parts in the fifth decimal place, about 0.001%. We have measured[19] over a broad temperature range (from 20 K to 750 K) in the infrared with about 0.01%.

Another very interesting method is that of Kramers and Kronig[20]. By measuring the normal-incidence reflectivity over a very wide spectral range, both optical constants can be ascertained. The two are related by the Hilbert transform. In theory, the entire spectrum must be measured, but in practice a broad range provides sufficient accuracy. The data so taken typically have an uncertainty of a few parts in the third decimal place. It is not as accurate as the prism techniques, but it provides information for samples with a higher absorption and over a wider spectral range.

13.11 RECAP

Measurement of the radiometric properties of materials covers a wide range of topics. There are many different types of materials—dielectrics, metals, smooth surfaces, rough surfaces, and even liquids and gases. There are many types of radiometric properties—reflectance, transmittance, emissivity, absorptance, and refractive index. There are many conditions—high and low values of the property, high and low values of the temperature, vacuum or gaseous environment, short wavelength or long, fast measurements or slow, and different required accuracies.

---

[17]Jenkins, F. A. and H. E. White, *Fundamentals of Optics*, McGraw Hill, 1955; M. Born and E. Wolf, *Principles of Optics*, Pergamon, 1959; E. Hecht and Zajak, *Optics*, Addison Wesley, 1974.

[18]Platt, B., H. Icenogle, J. Harvey, R. Korniski, and W. Wolfe, "Technique for measuring the refractive index and its change with temperature in the infrared," Journal of the Optical Society of America **65**, 1264, 1975.

[19]Wolfe, W., A. DeBell, and J. Palmer, "Status of cryogenic refractive index measurements," Proceedings of SPIE **245**, 163, 1980.

[20]Lang, M. and W. Wolfe, "Optical constants of fused silica and sapphire from 0.3 to 25 μm," Applied Optics **22**, 1267, 1983, and correction, ibid., 2949.

Thus, providing a cookbook of measuring techniques is a virtual impossibility. But this chapter is a beginning.

Emissivities are measured by energy conservation techniques and by comparison to blackbodies. Sometimes they can be a result of an absorptance measurement. Reflectivities divide into specular, hemispherical, and bidirectional, and techniques have been discussed for these—substitutional, VW, and Bennett-Koehler for specular; Coblentz, Gier-Dunkle, parabolic, and integrating spheres for hemispherical; and goniometric for bidirectional. External transmittance is measured in only one basic way, and absorption coefficient can sometimes be obtained from it. Absorption coefficient is also measured by the Stierwalt and calorimetric techniques. Refractive index is measured by minimum deviation and normal incidence methods as well as the Kramers-Kronig technique.

CHAPTER 14

RADIOMETRIC TEMPERATURES

Many radiometric measurements are aimed at determining the temperature of an object, not just its radiance[1,2]. Over the years several ways to do this with various approximations have been developed, and the procedures for obtaining a temperature from different properties of radiation have been developed, analyzed, and used.

The temperatures are defined. Approximate equations and exact equations are given. Then, for a reasonable set of conditions the differences among the several radiometric temperatures and the true temperature are plotted.

14.1 RADIOMETRIC TEMPERATURES

Several useful types have been defined[3,4]. These include *radiation temperature, brightness* or *radiance temperature, ratio temperature, distribution temperature, ratio-difference temperature* and *color temperature*. Of late, the somewhat ambiguous *effective temperature* has also been defined and used (in different ways), particularly with infrared simulators.

It will be seen that radiation and distribution temperatures are functions of the average emissivity of the body, while radiance and ratio temperatures are strong functions of the spectral emissivity and spectral atmospheric transmission and vary over the spectrum. Each is the temperature of a blackbody that gives the same radiometric properties as the body in question.

---

[1]Nerry F., J. Labed, and M. Stoll, "Emissivity signatures in the thermal infrared band for remote sensing: calibration procedure and method of measurement," Applied Optics **27**, 758, 1988.

[2] T. J. Quinn, *Temperature*, Academic Press, 1983.

[3] D. P. DeWitt, "Inferring Temperatures from Optical Radiation Measurements," Proceedings of SPIE **446,** 226, 1984.

[4] W. L. Wolfe, "Photometry and Radiometry," in *Physical Optics and Light Measurements,* D. Malacara, ed., Volume 20 of *Methods of Experimental Physics*, Academic Press, 1988.

14.1.1 RADIATION TEMPERATURE

Radiation temperature is the temperature of a blackbody that gives the same total radiance or radiant exitance as the body in question. The equality is

$$\sigma T_{rad}^4 = \varepsilon \sigma T^4, \tag{14-1}$$

where T_{rad} is the radiation temperature, T is the true temperature, and ε is the emissivity of the real body. Of course, the Stefan-Boltzmann constant σ cancels. Therefore,

$$T_{rad}^4 = \varepsilon T^4. \tag{14-2}$$

The emissivity must be the weighted average emissivity for all wavelengths or frequencies. The true temperature then is

$$T_t = \varepsilon^{-1/4} T_{rad}. \tag{14-3}$$

The relative error is given by

$$RE = 1 - \frac{T_{rad}}{T} = 1 - \varepsilon^{1/4}. \tag{14-4}$$

A more explicit representation that takes into account the spectral nature of the emissivity is

$$T_{rad}^4 = \int_0^\infty \varepsilon(\lambda) L_\lambda^{BB}(\lambda, T) d\lambda, \tag{14-5}$$

where L_λ^{BB} is the blackbody spectral radiance as a function of wavelength λ and temperature T. Although it has not been defined in the literature, one can define a photon temperature T_{ph} as

$$T_{ph}^3 = \varepsilon T^3. \tag{14-6}$$

14.1.2 RADIANCE TEMPERATURE

Radiance temperature is the temperature of a blackbody that has the same radiance or radiant exitance as the real body at a defined wavelength. The brightness temperature usually involves a measurement at about 0.7 μm.[5] The equality representing this definition is

$$\frac{c_1}{\pi\lambda^5(e^{c_2/\lambda T_b}-1)}=\frac{\varepsilon c_1}{\pi\lambda^5(e^{c_2/\lambda T}-1)},\tag{14-7}$$

where T_b is the radiance (brightness) temperature, c_1 is the first radiation constant, and c_2 is the second radiation constant. Simplification yields

$$e^{c_2/\lambda T}-1=\varepsilon(e^{c_2/\lambda T_b}-1).\tag{14-8}$$

For the Wien approximation for the Planck function, the ones drop out, and one has

$$T_b=\frac{T}{1-\frac{1}{x}\ln\varepsilon},\tag{14-9}$$

where x is $c_2/\lambda T$, the dimensionless frequency. If the approximation is not made, then one has

$$T_b=\frac{x}{\ln\left(1+\frac{e^x-1}{\varepsilon}\right)}T.\tag{14-10}$$

The relative errors in these two cases, respectively, are

$$RE=-\frac{1}{x}\ln\varepsilon,\tag{14-11}$$

$$RE=\frac{x}{\ln\left(1+\frac{e^x-1}{\varepsilon}\right)}.\tag{14-12}$$

---

[5] Lee, R. D. and E. Lewis, "Radiance Temperature at 6550 Å [655 nm] of the Graphite Arc," Applied Optics **5**, 1966.

14.1.3 RATIO TEMPERATURE

Ratio temperature is the temperature of a blackbody that has the same ratio of radiance or radiant exitance as the true body at two specific wavelengths. The defining equation is

$$\frac{c_1/\lambda_1^5(e^{c_2/\lambda_1 T_r}-1)}{c_1/\lambda_2^5(e^{c_2/\lambda_2 T_r}-1)} = \frac{\varepsilon_1 c_1/\lambda_1^5(e^{c_2/\lambda_1 T}-1)}{\varepsilon_2 c_1/\lambda_2^5(e^{c_2/\lambda_2 T}-1)},$$ (14-13)

where T_r is the ratio temperature. Simplification yields

$$\frac{(e^{c_2/\lambda_2 T_r}-1)}{(e^{c_2/\lambda_1 T_r}-1)} = \frac{\varepsilon_1(e^{c_2/\lambda_2 T}-1)}{\varepsilon_2(e^{c_2/\lambda_1 T}-1)}.$$ (14-14)

As with the radiance temperature, the Wien approximation can be used to show

$$T_r = \frac{T}{1-\frac{1}{X}\ln\varepsilon_r},$$ (14-15)

where

$$X = \frac{c_2}{\Lambda T},$$ (14-16)

and

$$\Lambda = \frac{\lambda_2\lambda_1}{\lambda_2-\lambda_1},$$ (14-17)

$$\varepsilon_r = \frac{\varepsilon_2}{\varepsilon_1}.$$ (14-18)

The form is the same, but the results are different. For radiance temperature, the error depends on the value of the emissivity at the wavelength of evaluation. For ratio temperature, it depends on the ratio of the emissivities at the two wavelengths. The relative error for the Wien approximation also has the same form. The Planck version of the error is not solvable in explicit form.

14.1.4 COLOR TEMPERATURE[6,7]

Color temperature is the temperature of a blackbody that has the same color coordinates on the chromaticity diagram as the real body. It explains the apparent paradox of why a blackbody at about 1200 °C looks orange and why the sun looks yellow. It has been used for visual measurements of apparent temperature, but is not as useful analytically as the other temperatures discussed in this chapter.

14.1.5 DISTRIBUTION TEMPERATURE

Distribution temperature is the temperature of a blackbody that has the same spectral distribution as the real body. In a sense it is a ratio temperature for an infinite number of wavelength pairs. The equation is (after elimination of constants)

$$\sum \frac{e^{c_2/\lambda_j T_d}-1}{e^{c_2/\lambda_j T_d}-1} = \sum \frac{\varepsilon_i(e^{c_2/\lambda_i T}-1)}{\varepsilon_j e^{c_2/\lambda_i T}-1)}, \qquad (14\text{-}19)$$

where T_d is the distribution temperature. The process is to reduce the mean-square difference between the two sets of ratios.

---

[6] Committee on Colorimetry, *The Science of Color,* Optical Society of America, T. Y. Crowell, 1963.

[7] J. S. Preston, *Introduction of the International Practical Temperature Scale 1968: some effects in relation to light sources, colour temperature, and colorimetry,* NBS Special Publication 300, Volume 7, Precision Measurement and Calibration; Radiometry and Photometry.

14.1.6 RATIO TEMPERATURE DIFFERENCE[8]

Ratio temperature difference has been used in medical imaging applications. It may be defined as the difference in temperature of two blackbodies that gives the same difference in temperature as two real bodies. In an imaging application, the two bodies can be considered to be two adjacent pixels. This is particularly useful in infrared medical imaging, where temperature patterns are the significant diagnostic. The thermodynamic temperature is not important; the areal pattern of temperature differences is important. Of course, one can then define differences in distribution, radiance, and radiation temperatures.

14.2 ATMOSPHERIC TRANSMISSION

It is clear in the development of these radiometric temperatures that the emissivity is a key element. In any field measurement the atmospheric transmission is also important. In such a field measurement the radiance is reduced by both the emissivity of the material and the transmission of the intervening atmosphere, and they act in equivalent ways for these calculations. Thus, in what appears above, the effects of atmospheric transmission can be incorporated by replacing the emissivity terms by the product of the emissivity and the transmission.

14.3 EFFECTIVE TEMPERATURE

Effective temperature is the temperature of a blackbody that gives the same irradiance at a plane as the body in question. This concept is often used in infrared simulation devices. The radiation from a set of resistors in a dynamic simulator is reduced by both the emissivity of the resistor and the proportion of the area it occupies. That is, a resistor array may have a fill factor η_{ff} of 60%. Then the resistors radiate in accordance with their temperature and emissivity, but the total radiation is "diluted" by the fact that 40% of the total area is at a different (lower) temperature with, perhaps, a different emissivity.

We can write an implicit expression for the effective temperature, assuming that all resistors are at the same, uniform temperature and that the background is uniform with a different temperature and emissivity. Then

---

[8] E. Dereniak, *Ratio Temperature Thermography*, Dissertation, The University of Arizona, Tucson, Arizona, 1976.

$$\int_{\lambda_1}^{\lambda_2} L_\lambda^{BB}(\lambda, T_{eff}) d\lambda = \eta_{ff} \int_{\lambda_1}^{\lambda_2} L_\lambda^{BB}(\lambda, T) d\lambda + (1 - \eta_{ff}) \int_{\lambda_1}^{\lambda_2} L_\lambda^{BB}(\lambda, T_{back}) d\lambda, \qquad (14\text{-}20)$$

where T_{eff} is the effective temperature and T_{back} is the background temperature. One can also include the transmission of the system in the same way as the transmission of the atmosphere. There is still some discrepancy among authors as to whether the transmission is part of the definition. However, this convenient fictional temperature allows one to evaluate fluxes on the test object rather nicely.

If the fill factor is considered simply as part of the overall efficiency of the radiation, then this effective temperature can be considered a special case of the radiance temperature. It is the temperature of a blackbody that gives the same radiation in a band as the body in question. The band is not monochromatic or quasi-monochromatic or even hemi-semi-demi-monochromatic. It is a band-effective temperature.

14.4 RECAP

The inference of a temperature from a radiometric measurement can be done with varying degrees of accuracy by the classic methods: *radiation temperature, ratio temperature, radiance* or *brightness temperature, distribution temperature*, and *color temperature*. Radiation and distribution temperature each use a measure of the emissivity over the entire spectrum. Ratio, radiance, and color temperatures use just a few points on the spectrum. The advantage of one over the other must be assessed in individual cases. Effective temperature means many things to many people and must be defined. It is usually the temperature that gives the best match to total radiation from a blackbody for the average radiation over an area.

CHAPTER 15

POLARIZATION EFFECTS

Every radiometer must be considered a polarizer to some extent, at least until the polarization properties have been shown to be insignificant. If it tends to induce polarization, it will also affect the polarization properties of any incoming polarized light. Then, the flux measured may not be representative of the flux emitted, because some has been rejected or altered by polarization. It is also a problem if the polarization properties of the calibration source are different from those of the unknown and the radiometers operate on these differences. This chapter describes some of the polarization properties of calibration sources and receivers and of radiometers and provides methods for calculating the polarization properties of an entire radiometer.

15.1 DESCRIPTIONS OF POLARIZATION

Light, that very special electromagnetic wave, can be polarized in different ways. The electric field may vibrate strictly in the vertical direction or strictly in the horizontal one, or in some other arbitrary plane. If it does, then it is called linear polarization in the direction of vibration, often specified as x and y or as s and p, standing for perpendicular *(senkrecht* in German) and parallel to the plane of incidence, respectively. The field vector, on the other hand, can follow the locus of a circle, and then the two possible components of polarization are left and right handed (circulation). The more modern descriptions are in terms of Mueller and Jones matrices. To some extent these two methods are complementary, but a full discussion of them is beyond the limits of this text. The Mueller method is described briefly and applied to some representative examples.

A column matrix is used to describe the beam of light in which the elements are {I M C S}. In this text, for ease of typing, the column vector is represented by curly brackets {} and a row vector by square brackets [ ]. The terms are the field intensity I with units of W m^{-2} (it seems useful to keep the symbols used by the writers in spite of the radiometric inconsistency); the degree of horizontal polarization, M; degree of $+45°$ polarization, C; and degree of right-circular polarization, S. For examples, {1 0 0 0} is an unpolarized beam of unit intensity; {1 -1 0 0} is a unit intensity vertically polarized beam. Table 15-1 provides information on most of the types and their Mueller representations.

Table 15-1. Polarization forms.	
Horizontal, linear	{1 1 0 0}
Vertical, linear	{1 -1 0 0]
+45, linear	{1 0 1 0}
-45, linear	{1 0 -1 0}
General linear	{1 cos2α sin2α 0 }
Right circular	{1 0 0 1}
Left circular	{1 0 0 -1}
Right, horizontal, elliptical	{1 0.6 0 0.8}
Right, vertical, elliptical	{1 -0.6 0 0.8}
Unpolarized	{1 0 0 0}

Devices are characterized by 4×4 Mueller matrices. A collection of some of these is in Shurcliffe[1]. The ideal, isotropic nonabsorbing plane parallel plate is represented by the unit matrix, which is ones on the diagonal and zeros elsewhere. An isotropic plane parallel plate with transmittance τ is represented by a diagonal matrix of value τ that multiplies the unit matrix. An ideal depolarizer is described by a matrix with a 1 in the first place in the matrix (first row, first column) and zeros elsewhere. An ideal linear horizontal polarizer has 0.5 in the first four positions of the matrix (rows 1 and 2 and columns 1 and 2) and zeros elsewhere. An ideal linear vertical polarizer has 0.5 and -0.5 in row 1 columns 1 and 2 and -0.5 and 0.5 in row 2 and columns 1 and 2, 0.5 in the corners. Shurcliffe also gives the general expressions. After a discussion of some of the polarization properties of radiometer components, we will return to these matrices.

---

[1]Shurcliffe, W. *Polarized Light: Production and Use*, Harvard University Press, 1962.

15.2 POLARIZATION FROM DIELECTRICS[2]

Plane parallel plates are used in some instruments as windows and as beamsplitters. Usually the windows are placed with their surfaces perpendicular to the incoming beam and beamsplitters are placed at 45 degrees. There are, of course, exceptions. Sometimes the windows are tilted a few degrees to avoid narcissus[3], and sometimes they are wedged a little to avoid interference effects. Sometimes beamsplitters are tilted at angles different from 45 degrees for various packaging reasons. Figure 15-1 shows the two components of polarization of the reflectivity as a function of angle of incidence, α, and for a nominal refractive index of 1.5. Figure 15-2 shows the same for the transmitted beam.

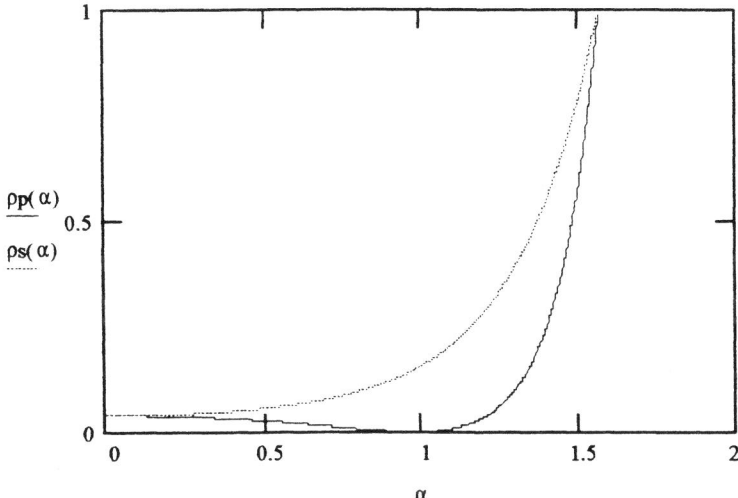

Figure 15-1. The polarization components of reflection from a dielectric.

---

[2]Jenkins, F. and H. White, *Fundamentals of Optics*, McGraw Hill, 1957.

[3]Wolfe, W. L., *Introduction to Infrared System Design*, SPIE Press, 1996.

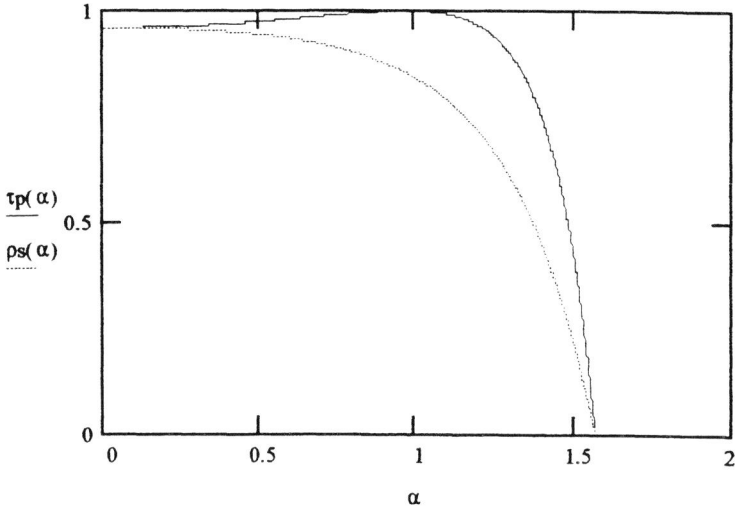

Figure 15-2. The polarization components of transmission from a dielectric.

It is clear from these figures that a plane, nonabsorbing, plane parallel plate is an almost ideal polarizer in the angular range from about 0.9 to 1.1 radians in both reflection and transmission. The so-called Brewster or polarizing angle at which the polarization is complete is given by the arctangent of the refractive index. It is also clear that there is no polarization introduced when the radiation is incident normally on the surface.

How can this affect a radiometer? The detector may be a photodetector with a smooth dielectric surface, or it may be a bolometer or thermocouple mounted in a vacuum behind an appropriate window. The light that is focused might very well be a convergent cone, characterized by some focal ratio. This means that the system is a partial polarization analyzer.

15.3 POLARIZATION FROM METALS[4]

Even metallic components have different reflectivities for different polarizations. It is more complicated in that elliptical polarization can be introduced. The reflectivities of metals (we do not need to know anything about the transmissivities of metals) are functions of the values of the real and imaginary components of the refractive index, n and k. Some examples are plotted in Figure 15-3.

---

[4]Born, M. And E. Wolf, *Principles of Optics*, Pergamon, 1959.

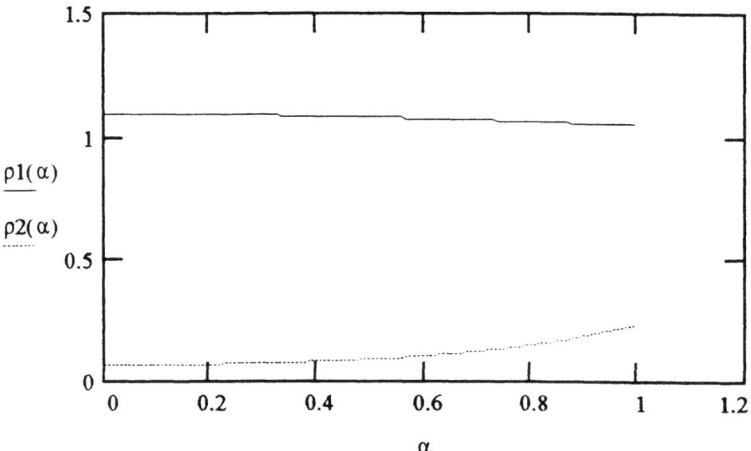

Figure 15-3. The perpendicular components of polarization for a metal with $n = 1.44$ and $k = 5.23$.

The expressions (if you really need them) are quite complicated and given at the end of this chapter. There is, in addition, a phase difference between the two components so that elliptical polarization is introduced. The polarization effects are evident even at normal incidence.

Many radiometers, if not most, have choppers that alternate the field of view from the scene to a calibration source. The light usually passes through openings in the chopper, but the calibration source gets polarized by the metallic chopper.

15.4 POLARIZATION FROM SURFACE SCATTERING

Surface scattering divides nicely into three types: scattering from a surface that has roughness small with respect to the wavelength, from a surface that has roughness large with respect to the wavelength, and surface particulates.

The components of polarization for the first of these were developed by a number of workers[5] and are shown in Figure 15-4 and 15-5 for metals and dielectrics,

---

[5]Maradudin, A. A. and D. L. Mills, "Scattering and absorption of electromagnetic radiation by a semi-infinite medium in the presence of roughness," Physical Review B **11**, 1392, 1975; Kröger, E. and E. Kretschmann, "Scattering of light by slightly rough surfaces or thin films including plasma resonance emission," Zeitschrift für Physik **237**, 1, 1970;

respectively. The nomenclature indicates the reflectivity for light polarized in the *s* or *p* directions and polarized in reflection in *s* or *p* directions as well. So, $ss(\theta)$ represents the reflectivity as a function of angle θ for incident light polarized perpendicular to the plane of incidence and measured with a polarizer after reflection in the perpendicular direction as well. The *sp* and *ps* components are essentially zero for all scatter angles. The *ss* component is almost one, and the *pp* component varies. It was assumed for this illustrative calculation that the angle of incidence is 0.1 radian and the measurement is made 0.1 radian out of the plane of incidence, the azimuthal direction. Increasing the angle of incidence will decrease the value of the *ss* component, but it will still be essentially constant.

Similar results are shown for dielectrics with a refractive index of about 1.5. The polarization is still apparent. All of these curves are normalized and will be very small with respect to specular reflective values.

The polarization for the surfaces that are rough with respect to the wavelength and oriented completely randomly is negligible. In fact, this is the kind of surface that is almost a complete depolarizer.

The components of polarization for the last of these were developed in detail by Spyak, who shows that the particles generate very little polarization; that is, the scattered light curve is about the same no matter what the polarization[6].

Although the scattering is often small, as usual, it must be shown that it is insignificant. This source of polarization occurs even at normal incidence.

Church, E. L., H. A. Jenkinson, and J. M. Zavada, "Relationship between Surface Scattering and Microtopographic Features," Optical Engineering **18**, 125, 1979; Videen, G., J. Hsu, W. Bickel, and W. Wolfe, "Polarized light scattered from rough surfaces," Journal of the Optical Society of America **9**, 1111, 1992.

[6]Spyak, P. R., *A cryogenic scatterometer and scatter from particulate contaminants on mirrors,* Dissertation, The University of Arizona, 1990.

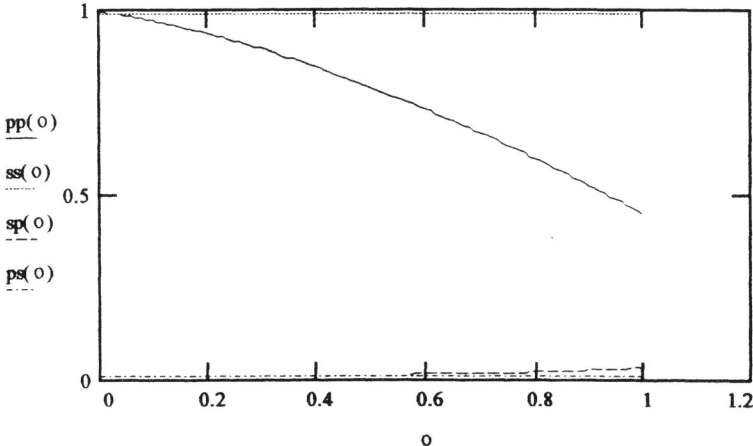

Figure 15-4. Polarization components for out-of-plane, small-roughness, surface scatter for metals.

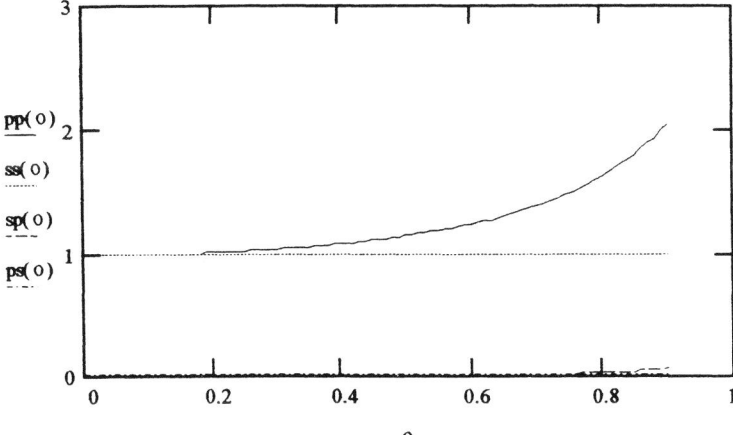

Figure 15-5. Polarization components for out-of-plane, small roughness, surface scatter for dielectrics.

I5.5 POLARIZATION FROM SLITS[7]

Partly as a result of the polarization aspects of diffraction, slits, such as those in spectrometers, can induce polarization. The electric vector tends to lie in the direction of the slit, and the polarization is greater for narrower slits and longer wavelengths. Metal grids are used as polarizers; they may be considered extensions of slits, or at least similar to slits. The difference seems to be that the direction of polarization is perpendicular to the long direction of the wires[8]. (This apparent contradiction is resolved by a careful consideration of the boundary-value conditions imposed on the perpendicular and parallel components of the electric field.) Diffraction gratings may also be thought of as similar to slits. They too have polarization properties in reflection[9] and in transmission, and so do echelette gratings[10].

I5.6 POLARIZATION AS A RESULT OF ATMOSPHERIC TRANSMISSION

It is well known that the light from the sky is polarized. Anyone doubting this should use his polarized sunglasses and rotate them while viewing the sky in the direction of the sun, opposite it, and to the left and right. The differences that should be observed are evidence of the polarization. (Not on an overcast day,

---

[7]Jones, R. V. and J. C. Richards, "Polarization of light by narrow slits," Proceedings of the Royal Society of London **A225**, 122, 1954; von Thiessen, G., "Polarisation des Lichtes beim Durchgang durch Metalle Spalte," Optik **2**, 266, 1947.

[8]Bird, G. R. and M. Parrish, "The wire grid as a near-infrared polarizer," Journal of the Optical Society of America **50**, 886, 1960; duBois, H. and H. Rubens, "Polarisation ungebeugter langswelliger Wärmstrahlen durch Drahtgitter," Annalen der Physik **35**, 243, 1911; Trentini, G. V., "Maximum transmission of electromagnetic waves by a pair of wire gratings," Journal of the Optical Society of America **45**, 883, 1955.

[9]Madden, R. P., "10 to 15 micron thick evaporated silver films for infrared gratings," Journal of the Optical Society of America, abstract, 1955.

[10]Meecham, W. C. And C. W. Peters, "Reflection of plane-polarized electromagnetic radiation from an echelette grating," Journal of Applied Physics **28,** 216, 1957.

please.) This effect is surely most serious when radiometric measurements are made in long paths out of doors. Calculations can be made based on the size distribution of the particulates and the wavelengths using the theories of Rayleigh and Mie[11].

15.7 REPRESENTATIVE MUELLER MATRICES

The Mueller matrix is a 4×4 matrix that multiplies the input column vector that represents the state of polarization of the input radiation to get the state of the output. Some of these are listed in the Shurcliffe book and are repeated and modified here.

The unit matrix, which represents vacuum, clean air, or an ideal, perfectly transparent, perpendicular plate, has ones on the diagonal and zeros elsewhere. If the plate has a transmittance, τ, less than one, the matrix is still the unit matrix but multiplied by τ. The totally absorbing plate has **all** zeros.
The unit matrix is

$$\begin{matrix} 1 & 0 & 0 & 0 \\ 0 & 1 & 0 & 0 \\ 0 & 0 & 1 & 0 \\ 0 & 0 & 0 & 1 \end{matrix}$$

The ideal depolarizer has a 1 in the first row, first column ($i=1, j=1$) position and zeros elsewhere. As Shurcliffe states, there is no such thing, but an integrating sphere and very rough plate are reasonable approximations. They would have something like a 1 in the first position and small values in the 1,2; 2,1; and 2,2 positions.

The ideal, horizontal, linear polarizer for horizontal polarization contains values of 0.5 in the first four positions and zeros elsewhere:

$$\begin{matrix} 0.5 & 0.5 & 0 & 0 \\ 0.5 & 0.5 & 0 & 0 \\ 0 & 0 & 0 & 0 \\ 0 & 0 & 0 & 0 \end{matrix}$$

---

[11]Van de Hulst, H. C., *Light Scattering by Small Particles*, Wiley, 1957; Bohren, C. F. and D. R. Huffman, *Absorption and Scattering of Light by Small Particles*, Wiley, 1983.

This could be the beamsplitter at near Brewster angle. When turned in the other direction, one gets ideal horizontal polarization:

$$
\begin{matrix}
0.5 & -0.5 & 0 & 0 \\
-0.5 & 0.5 & 0 & 0 \\
0 & 0 & 0 & 0 \\
0 & 0 & 0 & 0
\end{matrix}
$$

For +45 degree polarization the matrix is

$$
\begin{matrix}
0.5 & 0 & 0.5 & 0 \\
0 & 0 & 0 & 0 \\
0.5 & 0 & 0.5 & 0 \\
0 & 0 & 0 & 0
\end{matrix}
$$

For a partial linear polarizer then, a representative matrix for a tilted, plane parallel plate may be

$$
\begin{matrix}
0.5 & .1 & 0 & 0 \\
0.1 & .1 & 0 & 0 \\
0 & 0 & 0 & 0 \\
0 & 0 & 0 & 0
\end{matrix}
$$

Elliptical polarization is represented by a more complicated matrix (not surprisingly). This is the matrix that can be used to represent the polarization by a metal mirror or plate. The general elliptical-polarization ellipse is 0.5 times the following matrix:

$$
\begin{matrix}
1 & C_2Y & S_2Y & Z \\
C_2Y & C_2^2Y^2 & C_2S_2Y^2 & C_2YZ \\
S_2Y & C_2S_2Y^2 & S_2^2Y^2 & S_2YZ \\
Z & C_2YZ & S_2YZ & Z^2
\end{matrix}
$$

where

$$S_2=\sin(2\theta),\ C_2=\cos(2\theta),\ Y=\cos\left[2\arctan\left(\frac{b}{a}\right)\right],\ Z=\pm\sin\left[2\arctan\left(\frac{b}{a}\right)\right].\ (15\text{-}1)$$

The values of b, a, and θ must be determined from the optical constants of the metal.

The expressions for the reflectivities are, for the electric vector perpendicular to the plane of incidence,

$$\rho = \frac{(\cos\theta - u)^2 + v^2}{(\cos\theta + u)^2 + v^2},$$ (15-2)

$$2v^2 = -n^2(1 - \kappa^2) - \sin^2\theta + \sqrt{[n^2(1 - \kappa^2) - \sin^2\theta] + 4n^4\kappa^4},$$ (15-3)

$$2u^2 = n^2(1 - \kappa^2) - \sin^2\theta + \sqrt{[n^2(1 - \kappa^2) - \sin^2\theta] + 4n^4\kappa^4}.$$ (15-4)

For the TM mode, in which the electric field is parallel to the plane of incidence, the equations are

$$\rho^2 = \frac{[n^2(1 - \kappa^2)\cos\theta - u^2]^2 + [2n^2\kappa\cos\theta - v]^2}{[n^2(1 - \kappa^2)\cos\theta + u^2]^2 + [2n^2\kappa\cos\theta + v]^2},$$ (15-5)

$$\varphi = \arctan\left[2n^2\cos\theta \frac{2\kappa u - (1 - \kappa^2)v}{n^4(1 + \kappa^2)^2\cos^2\theta - (u^2 + v^2)}\right].$$ (15-6)

The reader is encouraged to participate in the gory details of these calculations!

15.8 RECAP

Radiometric measurements can be affected by the polarization properties of the instrument, a calibration source, the atmosphere, and the unknown. Every radiometer must be regarded as a polarization analyzer. Metallic chopper blades can induce elliptical polarization; canted beamsplitters can polarize the incoming light; spectrometer slits can induce polarization as can surface and atmospheric scattering. Mueller matrices can be used to evaluate the overall polarization characteristics of any radiometer if the component matrices are known. Alternatively, the overall polarization properties can be measured by a series of tests using linear and circular polarizers and quarter-wave plates.

Appendix
Some Geometric Configuration Factors

Introduction

In this section we consider several different geometries from the point of view of throughput, configuration factors and fractional areas, and with extensions.

Preliminaries and Definitions[1]

The fundamental equation of radiative transfer has been developed earlier; it is

$$dP = \frac{L \, dA_1 \cos\theta_1 \, dA_2 \cos\theta_2}{\rho^2},$$ (1)

where the line-of-sight distance is indicated by ρ to emphasize that it is a variable. The throughput, indicated by Z, is given by the multiplier of the radiance, i.e.,

$$Z = \frac{dA_1 \cos\theta_1 \, dA_2 \cos\theta_2}{\rho^2}.$$ (2)

For a lambertian source this becomes

$$E_{dA_1} = \frac{M}{\pi} \left[\frac{\cos\theta_1 \, dA_2 \cos\theta_2}{\rho^2} \right] = M \left[\frac{\cos\theta_1 \, dA_2 \cos\theta_2}{\pi\rho^2} \right].$$ (3)

The geometric configuration factor, also called the view factor, interaction factor, angle factor, etc., is given by

$$GCF = \frac{Z}{\pi} = \frac{dA_1 \cos\theta_1 \, dA_2 \cos\theta_2}{\pi\rho^2}.$$ (4)

The fractional area F is the GCF divided by the area. Of course, the question then is, "What area?" The answer then is, "Whatever area you want!" Then the area must be indicated. The clearest way to explain this is by example:

---

[1] A. Sarofim and H. Hottel, *Radiative Transfer*, McGraw Hill, 1967.

$$F_{dA_1 - A_2} = \frac{GCF}{dA_1} = \frac{dA_1 \cos\theta_1 dA_2 \cos\theta_2}{\pi \rho^2 dA_1} = \frac{\cos\theta_1 dA_2 \cos\theta_2}{\pi \rho^2}. \qquad (5)$$

In a similar way the fractional area between two finite areas is indicated by

$$F_{12} = F_{A_1 - A_2} = \frac{GCF}{A_1}. \qquad (6)$$

PLANE PARALLEL RECTANGLES

Two versions exist. One is between a differential element and a plane; the other is between two finite planes.

Differential Element and a Parallel Rectangle

Consider radiation from a differential element of area dA_1 parallel to a rectangular plate of dimensions 0 to X and 0 to Y a distance Z away, as shown in Figure 1.

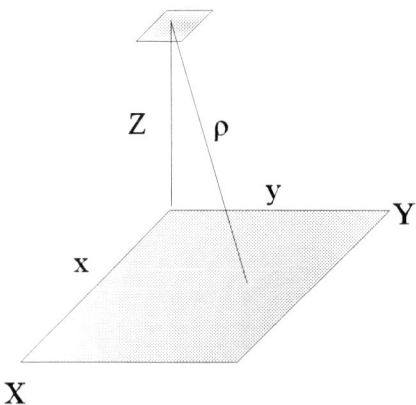

Figure 1. Differential element and parallel plane.

The fundamental equation of transfer says

$$dP = \frac{L\,dA_1\cos\theta_1\,dA_2\cos\theta_2}{\rho^2}.$$ (7)

The irradiance on dA_1 may be calculated by dividing through by dA_1; the two cosines with respect to the normals are both the same, by virtue of alternate interior angles. The elementary area is $dx\,dy$, and the distance is given by the three-dimensional pythagorean theorem:

$$dE = \frac{L\cos^2\theta\,dx\,dy}{x^2+y^2+Z^2} = \frac{LZ^2\,dx\,dy}{\rho^4},$$ (8)

where Z is a fixed distance, since these are two parallel plates a given distance apart. The angle between the normal and the surface is always just $\cos\theta$. There is no azimuthal angle to include. The cosine is just the z height divided by the distance on the plane. Thus

$$dE = \frac{LZ^2\,dx\,dy}{(x^2+y^2+Z^2)^2}.$$ (9)

The result is

$$E = Z^2 \int_0^X \int_0^Y \frac{L\,dx\,dy}{(x^2+y^2+Z^2)^2}.$$ (10)

It is interesting that the cosine series for the radiance can be used to represent nonisotropic surfaces. In this case each cosine increases the power of Z by one, and increases the power of the denominator by one.

When the radiance is eliminated from the equation, the expression is for the fractional area; that is,

$$F_{dA_1-A_2} = Z^2 \int_0^X \int_0^Y \frac{dx\,dy}{(x^2+y^2+Z^2)^2}.$$ (11)

The *GCF* for this geometry is

$$GCF = Z^2 \int_0^X \int_0^Y \frac{dxdydA_1}{(x^2 + y^2 + Z^2)^2}. \tag{12}$$

Two Finite Parallel Rectangles

The extension to two finite rectangles is simple; integration must be made over the upper plane instead of just using the differential element. The geometry is shown in Figure 2. The distance between differential elements of area is given by

$$\rho^2 = (x_2 - x_1)^2 + (y_2 - y_1)^2 + Z^2. \tag{13}$$

This expression for ρ needs to put in the denominator and integration carried out over both plates:

$$E = Z^2 L \int\int\int\int \frac{dx_1 dy_1 dx_2 dy_2}{[(x_2 - x_1)^2 + (y_2 - y_1)^2 + Z^2]^2}. \tag{14}$$

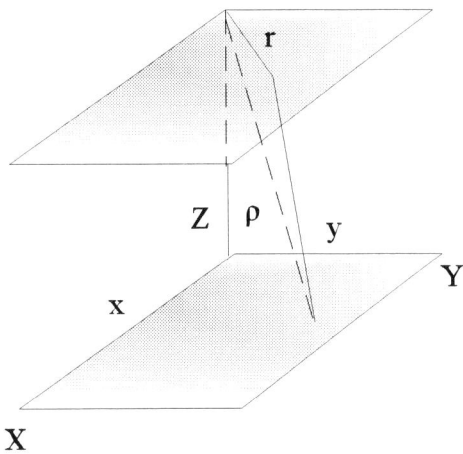

Figure 2. Plane parallel plates.

The result for a rectangle and a differential element that is also parallel to it and arises from the center of the rectangle can be obtained by a little consideration. The integration for the case just considered was from 0 to X and 0 to Y. For the centered rectangle, the integration is from $-X$ to X and $-Y$ to Y. Thus we have

$$\Omega'_{dA_1-A_2} = \frac{X}{(X^2+Y^2)^{1/2}}\sin^{-1}\frac{Y}{(X^2+Y^2+Z^2)^{1/2}} + \frac{Y}{(Z^2+Y^2)^{1/2}}\sin^{-1}\frac{X}{(X^2+Y^2+Z^2)^{1/2}} \cdot \quad (15)$$

The same tricks with the cosine expansion may be applied here, of course, but the complexity increases dramatically.

TWO PERPENDICULAR PLANE RECTANGLES

First consider the differential element that is arranged perpendicular to a plane plate. Now the differential element is assumed to lie in a plane that is perpendicular to the plate and the plane is at one edge of the plate.

Then the distance ρ is still the same, except that z is variable:

$$\rho^2 = (x_2-x_1)^2 + y^2 + z^2. \quad (16)$$

The cosine of the angle between the horizontal plate and its normal is the same:

$$\cos\theta_1 = \frac{z}{\rho}. \quad (17)$$

The cosine of the angle between the line of centers and the vertical differential element is

$$\cos\theta_2 = \frac{\sqrt{(x-\xi)^2+y^2}}{\rho}. \quad (18)$$

Therefore

$$E = \int\int \frac{Lz\sqrt{(x-\xi)^2+\zeta^2}dxdy}{\rho^4}. \quad (19)$$

Now consider a horizontal plate that meets a vertical plate along the x axis, as shown in Figure 3:

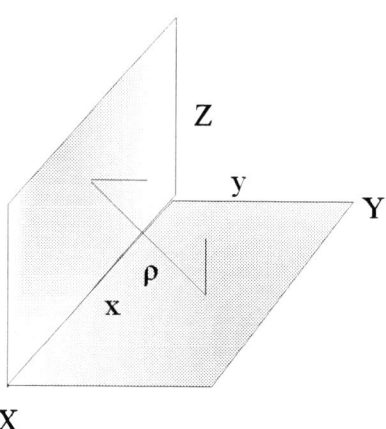

Figure 3. Perpendicular planes in edge contact.

$$\rho^2 = (x_2 - x_1)^2 + y_2^2 + z^2, \tag{20}$$

$$\cos\theta_1 = \frac{y_2}{\rho}, \tag{21}$$

$$\cos\theta_2 = \frac{\sqrt{(x_2 - x_1)^2 + y_2^2}}{\rho}. \tag{22}$$

Therefore

$$P = L \iiiint \frac{z\sqrt{(x_2 - x_1)^2 + y_2^2}\, dx_1 dx_2 dy dz}{\rho^4}. \tag{23}$$

Remember that x, y, z, ζ, ξ, and ρ are all variables.

Contour Integration

The point was made that the projected solid angle, or πGCF, is a function only of the periphery of the area enclosed. Thus, the idea of contour integration raises its delightful head. The projected solid angle is the area in question projected onto the unit sphere. This area forms a conical volume with an apex at the center of the sphere. This is the projected solid angle. The vector area, the area with an outward-drawn normal, is defined. Then

$$d\Omega' = -d\vec{A}. \qquad (24)$$

The integral of this surface area over the entire volume is equal to zero so that

$$\int_{sphere} d\vec{A} = -\int_{cone\,sides} d\vec{A}. \qquad (25)$$

The area of the cone sides is

$$\int d\Omega' = \oint d\gamma/2. \qquad (26)$$

This wonderfully simple result can be applied to the rectangle problem we have considered before, the rectangle and the differential element on the line at the corner of the rectangle and parallel to the plane. The γ's are

$$\gamma_1 = \sin^{-1}\frac{X}{(X^2+Z^2)^{1/2}}, \qquad (27)$$

$$\gamma_2 = \sin^{-1}\frac{Y}{(X^2+Y^2+Z^2)^{1/2}}, \qquad (28)$$

$$\gamma_3 = \sin^{-1}\frac{X}{(X^2+Y^2+Z^2)^{1/2}}, \qquad (29)$$

$$\gamma_4 = \sin^{-1}\frac{Y}{(Y^2+Z^2)^{1/2}}. \qquad (30)$$

There remains the task of evaluating the projections of these vectors. Note that, based on the geometry, each of the outward drawn normals is perpendicular to one of the planes shown in the figure:

$$\Omega' = \frac{1}{2}\left|\gamma_1\cos\frac{\pi}{2} + \gamma_2\cos(\frac{\pi}{2}-\gamma_1) + \gamma_3\cos(\frac{\pi}{2}-\gamma_4) + \gamma_4\cos\frac{\pi}{2}\right|, \tag{31}$$

$$\Omega' = \frac{1}{2}\left|\gamma_2\sin\gamma_1 + \gamma_3\sin\gamma_4\right|, \tag{32}$$

$$\Omega' = \frac{1}{2}\left|\frac{X}{(X^2+Z^2)^{1/2}}\sin^{-1}\frac{Y}{\rho} + \frac{Y}{(Y^2+Z^2)^{1/2}}\sin^{-1}\frac{Y}{\rho}\right|. \tag{33}$$

The angles are measured with respect to the normal to the differential area, in this case the z axis. This is the same answer obtained before by brute force and tedious integration.

The second example can be obtained with equal ease. Consider the same problem, except that the differential element is perpendicular to the rectangular surface. Then only the projections need to be considered; the values of the γ's are all the same. Now we can recognize that the cosines of the angles are, respectively, 1, $\pi/2 - \gamma_3$, γ_4, and $\pi/2$.

The contour technique can be applied to two finite surfaces. A little more maneuvering is necessary. The double integral is now over both surface areas, but one is the closed integral over the periphery:

$$\Omega' = \int_{A_1} \frac{1}{2}\oint d\vec{\gamma}\cdot d\vec{A}. \tag{34}$$

The vector $d\gamma$ is given by

$$d\vec{\gamma} = \frac{\vec{r}X d\vec{l}}{r^2}, \tag{35}$$

where r is the vector from dA_1 to dA_2 and l is a vector tangent to A_2 and equal in magnitude to a small differential on the perimeter of A_2. Then

$$\nabla \ln r = \frac{\hat{r}}{r} = \frac{1}{r}\frac{\vec{r}}{r} = \frac{\vec{r}}{r^2}. \tag{36}$$

Then

$$d\vec{\gamma} = \nabla X(\ln r \, d\vec{l}) \tag{37}$$

and

$$\Omega' = \frac{1}{2}\int_{A_1}\oint \nabla X(\ln r \, d\vec{l}) \cdot d\vec{A}_1. \tag{38}$$

Stokes theorem can be applied:

$$\Omega' = \frac{1}{2}\oint\oint \ln r \cos\theta \, dl \, dl'. \tag{39}$$

and

$$\cos\theta = \cos\theta_x \cos\theta'_x + \cos\theta_y \cos\theta'_y + \cos\theta_z \cos\theta'_x. \tag{40}$$

The right-hand-screw rule applies.

One example is perpendicular planes, as shown. The perimeter l_1 is in the xy plane and is parallel to either the x axis or the y axis. The perimeter of l_2 is in the yz plane and is parallel to either the x axis or the z axis. Therefore dl_1 is perpendicular to dl_2 except when they are both parallel to the x axis, and then they are parallel. Therefore $\cos\theta$ is one when they are parallel and zero otherwise. The contour integral is therefore

$$\Omega' = \sum \int\int \ln r \, dx_1 \, dx_2. \tag{41}$$

The summation is over each of the four edges. The sign is positive for the right-hand screw. Therefore the integral is

$$\Omega' = \frac{1}{2}\int\int \ln[(x_1 - x_2)^2 + b^2]^{1/2} dx_1 \, dx_2, \tag{42}$$

where b is a little different for each of the integrations. The integral can be simplified because it is of a particular form; that is, a double integral can be reduced to a single one as follows:

$$\int \int f[(x_1 - x_2)^n + b^2] dx_1 dx_2 = 2 \int (1 - y) f(y^n, c) dy, \qquad (43)$$

for even values of n and for y being the difference $x_1 - x_2$. Then

$$\Omega' = \int (1 - y) \ln(y^2 + b^2)^{1/2} dy. \qquad (44)$$

The values of b are 0, Y^2, Z^2, and $Y^2 + Z^2$ for the successive positions of dx_1 and dx_2 on the y axis and the z axis, respectively: $0,0$: $Y,0$: $0,Z$; and Y,Z. Integration gives

$$\Omega' = \frac{1}{2} \left| -\frac{3}{2} + 2b \tan^{-1} \frac{1}{b} + \frac{1}{2} \ln \frac{(b^2)^{b^2}}{(1 + b^2)^{b^2 - 1}} \right|. \qquad (45)$$

Then evaluation provides

$$\Omega' = \left[XY \tan^{-1} \frac{X}{Y} + XZ \tan^{-1} \frac{X}{Z} - X(Y^2 + Z^2)^{1/2} \tan^{-1} \frac{X}{(Y^2 + Z^2)^{1/2}} + \right.$$

$$\left. \frac{1}{4} \ln \left(\frac{Y^{2^{Y^2}}}{(X^2 + Y^2)^{Y^2 - X^2}} \frac{Z^{2^{Z^2}}}{(X^2 + Z^2)^{Z^2 - X^2}} \frac{(X^2 + Y^2 + Z^2)^{Z^2 + Y^2 - X^2}}{(Y^2 + Z^2)^{Y^2 + Z^2}} \right) \right]. \qquad (46)$$

These examples have all been for direct transfer from one surface to another; it has been assumed that there is no multiple interaction between the surfaces. Now that assumption will be relaxed, and several geometries will be considered in which there are many, even an infinite number of, interactions.

INFINITE PLANES, CONCENTRIC SPHERES, AND INFINITELY LONG CYLINDERS

The Direct Method

A direct method may be applied to two infinite planes that have non-zero values of reflectivity. The irradiance at the left plane, plane 1, is given by the sum of the radiation that is emitted from the left plane and multiply reflected by the two planes, plus the radiation that is emitted from the right plane and is also multiply reflected. The two planes are infinite and therefore of the same area. The irradiance on the first (the left) is

$$E_1 = \varepsilon_1 M_1 \rho_2 (1 + \rho_1 \rho_2 + \rho_1^2 \rho_2^2 + \cdots) + \varepsilon_2 (1 + \rho_1 \rho_2 + \rho_1^2 \rho_2^2 + \cdots). \tag{47}$$

Using the expression for the sum of an infinite geometric series, one has

$$E_1 = \frac{\varepsilon_1 \rho_2 M_1}{1 - \rho_1 \rho_2} + \frac{\varepsilon_2 M_2}{1 - \rho_1 \rho_2}. \tag{48}$$

Then, using the fact that for an opaque sample $\rho = 1 - \varepsilon$ (see Chapter 3 for specific conditions under which this is valid), one has

$$E_1 = \frac{\dfrac{M_1}{\varepsilon_2} + \dfrac{M_2}{\varepsilon_1} - M_1}{\dfrac{1}{\varepsilon_1} + \dfrac{1}{\varepsilon_2} - 1}. \tag{49}$$

The Total Radiation Method

Another way to perform this calculation is what may be called the total radiation method. In this, the total flux density M exiting a surface is composed of the reflected irradiance ρE and the emitted flux density εM^{BB}.

Therefore

$$M = \varepsilon M^{BB} + \rho E \tag{50}$$

and

$$E = \frac{M - \varepsilon M^{BB}}{\rho}. \tag{51}$$

Let us apply this total radiation method to infinite parallel planes. Then the irradiance on plane 1 is given by

$$E_1 = \frac{M_1 - \varepsilon_1 M_1^{BB}}{\rho_1} = \int_{A_2} M_2 dF_{12} + \int_{A_1} M_1 dF_{11} \tag{52}$$

$$= M_2 F_{12} + M_1 F_{11} = M_2. \tag{53}$$

Here F_{ij} is the fractional area of radiation. It is the area of the receiver divided by the total area the radiator can radiate to, discussed above. Then, since F_{11} is zero and F_{12} is one (the surface cannot see itself, and the infinite plane is all it can see), one has

$$\frac{M_1 - \varepsilon_1 M_1^{BB}}{\rho_1} = M_2. \tag{54}$$

Similarly

$$\frac{M_2 - \varepsilon_2 M_2^{BB}}{\rho_1} = M_1. \tag{55}$$

Then

$$\frac{M_1 - \varepsilon_1 M_1^{BB}}{\rho_1} = M_1 \rho_2 + \varepsilon_2 E_2^{BB} \tag{56}$$

and

$$M_1 = \frac{\rho_1 \varepsilon_2 M_2^{BB} + \varepsilon_1 M_1^{BB}}{1 - \rho_1 \rho_2}. \tag{57}$$

Then by algebraic manipulation it can be found that

$$M_1 = \frac{\dfrac{M_1^{BB}}{\varepsilon_2} + \dfrac{M_2^{BB}}{\varepsilon_1} - M_2^{BB}}{\dfrac{1}{\varepsilon_1} + \dfrac{1}{\varepsilon_2} - 1} . \qquad (58)$$

The irradiance then is

$$E_1 = \frac{\dfrac{M_1^{BB}}{\varepsilon_2} + \dfrac{M_2^{BB}}{\varepsilon_1} - M_1^{BB}}{\dfrac{1}{\varepsilon_1} + \dfrac{1}{\varepsilon_2} - 1} . \qquad (59)$$

The net energy exchange is

$$\Delta M = \frac{M_1^{BB} - M_2^{BB}}{\dfrac{1}{\varepsilon_1} + \dfrac{1}{\varepsilon_2} - 1} . \qquad (60)$$

Concentric Spheres

The inside sphere has subscripts 1; the outside sphere, 2. The basic transfer equation in this new form is

$$\frac{E_2 - \varepsilon_2 E_2^{\,e}}{\rho_2} = \sum_i \int_{A_i} E_i dF_{2i} = \int_{A_1} E_1 dF_{21} + \int_{A_2} E_2 dF_{22}. \qquad (61)$$

The fractional area $F_{12} = 1$ because A_2 completely surrounds A_1, and, since the area A can *see* its own area except where it is shielded by the area A_1, then $F_{21} = A_1/A_2$ and $F_{22} = 1 - A_1/A_2$, so that

$$\frac{E_2 - \varepsilon_2 E_2^{\,e}}{\rho_2} = \frac{E_1 A_1}{A_2} + E_2 \left(1 - \frac{A_1}{A_2} \right). \qquad (62)$$

Similarly

$$\frac{E_1 - \varepsilon_1 E_1^{\,e}}{\rho_1} = \int_{A_1} E_1 dF_{11} + \int_{A_2} E_2 dF_{12} = E_2 F_{12} = E_2. \tag{63}$$

Then

$$E_1 = \frac{A_1 \varepsilon_1 (1 - \varepsilon_2) E_1^{\,e} + A_2 \varepsilon_2 [E_2^{\,e} + \varepsilon_1 (E_1^{\,e} - E_2^{\,e})]}{A_2 \varepsilon_2 + A_1 \varepsilon_1 (1 - \varepsilon_2)}. \tag{64}$$

Then the net flux density is given by

$$\Delta P = \frac{E_1^{\,e} - E_2^{\,e}}{\dfrac{1}{\varepsilon_1} + \dfrac{A_1}{A_2}\left(\dfrac{1}{\varepsilon_2} - 1\right)}. \tag{65}$$

It will now be stated without proof that for mixed diffuse and specular spheres and cylinders, the GCF is given by[1]

$$\frac{GCF}{A_1} = \frac{1}{\dfrac{1}{\varepsilon_1} + \dfrac{\rho_{s,2}}{1 - \rho_{s,2}} + \dfrac{A_1}{A_2}\dfrac{\rho_{d,2}}{\varepsilon_2}\dfrac{1}{1 - \rho_{s,2}}}, \tag{66}$$

where the d subscript indicates diffuse reflectivity and s indicates specular reflectivity. In this model the diffuse reflectivity is the hemispherical reflectivity of an isotropic surface. Confidence in the model is obtained by investigating the two limits, purely diffuse and purely specular. In the first case, the specular reflectivities are zero and

$$\frac{GCF}{A_1} = \frac{1}{\dfrac{1}{\varepsilon_1} + \dfrac{A_1}{A_2}\dfrac{\rho_{d,2}}{\varepsilon_2}} = \frac{1}{\dfrac{1}{\varepsilon_1} + \dfrac{A_1}{A_2}\left(\dfrac{1}{\varepsilon_2} - 1\right)}. \tag{67}$$

This does agree with the previous result. Similarly, when purely specular surfaces are assumed, the result is the same as for planes, as it should be. There are an infinite number of reflections.

SOME OTHER GEOMETRIES

Parallel Circular Plates

Consider a differential area on axis parallel to a circular disk. Then $\rho^2 = r^2 + Z^2$ and the cosines are equal to Z/ρ. Therefore, by substitution and straightforward integration,

$$dE = LZ^2 \int_0^{2\pi} \int_0^r \frac{r \, dr \, d\varphi}{\rho^4} = \pi L Z^2 \int \frac{2r \, dr}{(Z^2 + r^2)^2}, \qquad (68)$$

$$dE = \frac{\pi L Z^2}{Z^2 + r^2}. \qquad (69)$$

This, fortunately, is the same result as obtained in Chapter 1 using integration via trigonometric functions, but here it gives an entree to the method with circular geometries.

Consider the interchange between two circular parallel plates. We can write

$$\rho^2 = Z^2 + (r_2 \cos\varphi_2 - r_1 \cos\varphi_1)^2 + (r_2 \sin\varphi_2 - r_1 \sin\varphi_1)^2, \qquad (70)$$

$$\cos\theta_1 = \cos\theta_2 = \frac{Z}{\rho}, \qquad (71)$$

$$dE = LZ^2 \int_0^{2\pi} \int_0^{2\pi} \int_0^{R_1} \int_0^{R_2} \frac{r_1 \, dr_1 \, r_2 \, dr_2 \, d\varphi_1 \, d\varphi_2}{\left[Z^2 + (r_2 \cos\varphi_2 - r_1 \cos\varphi_1)^2 + (r_2 \sin\varphi_2 - r_1 \sin\varphi_1)^2 \right]^2} \qquad (72)$$

$$dE = 4\pi^2 LZ^2 \int_0^{R_1} \int_0^{R_2} \frac{r_1 \, dr_1 \, r_2 \, dr_2}{\left[Z^2 + (r_2 \cos\varphi_2 - r_1 \cos\varphi_1)^2 + (r_2 \sin\varphi_2 - r_1 \sin\varphi_1)^2 \right]^2}. \qquad (73)$$

Perpendicular Circular Plates

We can learn from the treatment of the perpendicular plates. The geometry is the same, but the boundaries are circular. Therefore, if we take the expressions from the rectangle case and translate into cylindrical coordinates, we are there. The distance ρ is still the same, except that z is variable. We will use x_1, x_2, y_1, y_2 for the cartesian coordinates and similarly use subscripts 1 and 2 on r and φ, the azimuthal angle. Then

$$x_i = r_i \cos\theta_i \qquad z_i = r_i \sin\theta_i. \tag{74}$$

Therefore

$$\rho^2 = (x_2 - x_1)^2 + y_2^2 + z^2 = (r_2 \cos\varphi_2 - r_1 \cos\varphi_1)^2 + r_2^2 \sin^2\varphi_2 + z^2. \tag{75}$$

The cosine of the angle between the horizontal plate and its normal is the same:

$$\cos\theta_1 = \frac{z}{\rho}. \tag{76}$$

The cosine of the angle between the line of centers and the vertical differential element is

$$\cos\theta_2 = \frac{\sqrt{(x_2 - x_1)^2 + y_2^2}}{\rho} = \frac{\sqrt{(r_2 \cos\varphi_2 - r_1 \cos\varphi_1)^2 + y_2^2}}{\rho}. \tag{77}$$

Plug this all in, as usual:

$$dE = L \int\int\int\int \frac{r_1 \sin\varphi_1 \sqrt{(r_1 \cos\varphi_1 - r_2 \cos\varphi_2)^2 + r_2^2 \sin^2\varphi_2} \, r_1 r_2 \, dr_1 \, dr_2 \, d\varphi_1 \, d\varphi_2}{\left[(x_2 - x_1)^2 + y^2 + z^2\right]^2}. \tag{78}$$

Wow!

SUMMARY AND OVERVIEW

Several techniques have been given for the evaluation of the radiation interchange between a variety of surface shapes. There are vast tables of view factors or *GCF*'s and other texts go into considerably more detail for these evaluations.

There is an algebra for view factors that allows one to infer some results from the combinations of other results. There are summation rules and symmetry rules and more. The reader is referred there.

It is my belief that modern computing techniques will take care of most of the problems in one of several ways. There are essentially three types of integration that will take care of most problems. These are the integration over rectangular areas, and the reader should have the forms for both parallel and perpendicular arrangements clearly in mind. There are circular arrangements, and the same is true. Note that a circle to a rectangle can be handled using appropriate coordinate systems on each. The circles can be stretched to ellipses. Finally, when all else fails, the finite element approach works reasonably well—if the differential element is taken small enough.

Consider making your own table for the following integrands: parallel rectangles, perpendicular rectangles, parallel circles, perpendicular circles, parallel ellipses, perpendicular ellipses, parallel circle and rectangle, parallel circle and ellipse, perpendicular circle and rectangle, perpendicular circle and ellipse. With proper integrands in the correct coordinate system it is easy to write computer programs to perform the integration.

Index

WILLIAM L. WOLFE was born in Yonkers, New York, at a very early age. He received a BS in physics, cum laude, from Bucknell University. He did graduate work at the University of Michigan, where he received an MS in physics and an MSE in electrical engineering. (The reception of these degrees was not automatic, but required a certain amount of work.) While attending the University of Michigan, he held the positions of Research Engineer and Lecturer, and engaged in projects such as the development of a full-body thermographic scanner for medical analyses and a hot-rolled strip steel defect detector. In 1966 he finally left school to join the Honeywell Radiation Center in Lexington, MA, as Department Manager and Chief Engineer. While at Honeywell he supervised the development of infrared tank night-driving systems, a radiometer for sensing the infrared horizon from orbit, and infrared rifle sights. In 1969 he returned to school, specifically the University of Arizona, where he became Professor of Optical Sciences in the Optical Sciences Center. While there he supervised over 30 students, developed a cryogenic refractometer and the first automated scatterometer for three-dimensional scatter measurements, a probe that measured the solar flux in the atmosphere of Venus, a helicopter night-driving system, and other devices. In 1996 he became (officially) Professor Meritless, and under this guise has been investigating optical cancer detection and the early measurement of glaucoma. He has been a Fellow and on the Board of Directors of the Optical Society of America; a Senior Member of IEEE; and a Fellow, Life Member, and past president of SPIE—The International Society for Optical Engineering. He is the Editor-in-Chief of *Infrared Physics and Technology,* coeditor of *The Infrared Handbook*, Associate Editor of the *Handbook of Optics,* and author of Tutorial Texts on *Infrared System Design* and *Imaging Spectrometers*. He is the proud father of three wonderful children, who are no longer teenagers, two grandsons, and a granddaughter. In his spare time, he sings, fly fishes, gardens, and uses his wife's phone.